AF179758

Piezoelectric Actuators - Principles, Design, Experiments and Applications

Edited by Hu Huang and Jianping Li

Published in London, United Kingdom

IntechOpen

Supporting open minds since 2005

Piezoelectric Actuators - Principles, Design, Experiments and Applications
http://dx.doi.org/10.5772/intechopen.92533
Edited by Hu Huang and Jianping Li

Contributors
Joel Francis Shields, Edward Konefat, Rashed A. Adnan Islam, José Carlos de Carvalho Pereira, Tinghai Cheng, Xiaosong Zhang, Xiaohui Lu, Hengyu Li, Qi Gao, Guangda Qiao, Tian-Bing Xu, Sahil P. Wankhede, Zhonghua Zhang, Junwu Kan, Song Chen, Jianping Li, Jianming Wen, Yili Hu, Nen Wan, Lidong He, Lin Zhang, Hu Huang

© The Editor(s) and the Author(s) 2021
The rights of the editor(s) and the author(s) have been asserted in accordance with the Copyright, Designs and Patents Act 1988. All rights to the book as a whole are reserved by INTECHOPEN LIMITED. The book as a whole (compilation) cannot be reproduced, distributed or used for commercial or non-commercial purposes without INTECHOPEN LIMITED's written permission. Enquiries concerning the use of the book should be directed to INTECHOPEN LIMITED rights and permissions department (permissions@intechopen.com).
Violations are liable to prosecution under the governing Copyright Law.

CC BY

Individual chapters of this publication are distributed under the terms of the Creative Commons Attribution 3.0 Unported License which permits commercial use, distribution and reproduction of the individual chapters, provided the original author(s) and source publication are appropriately acknowledged. If so indicated, certain images may not be included under the Creative Commons license. In such cases users will need to obtain permission from the license holder to reproduce the material. More details and guidelines concerning content reuse and adaptation can be found at http://www.intechopen.com/copyright-policy.html.

Notice
Statements and opinions expressed in the chapters are these of the individual contributors and not necessarily those of the editors or publisher. No responsibility is accepted for the accuracy of information contained in the published chapters. The publisher assumes no responsibility for any damage or injury to persons or property arising out of the use of any materials, instructions, methods or ideas contained in the book.

First published in London, United Kingdom, 2021 by IntechOpen
IntechOpen is the global imprint of INTECHOPEN LIMITED, registered in England and Wales, registration number: 11086078, 5 Princes Gate Court, London, SW7 2QJ, United Kingdom
Printed in Croatia

British Library Cataloguing-in-Publication Data
A catalogue record for this book is available from the British Library

Additional hard and PDF copies can be obtained from orders@intechopen.com

Piezoelectric Actuators - Principles, Design, Experiments and Applications
Edited by Hu Huang and Jianping Li
p. cm.
Print ISBN 978-1-83968-831-7
Online ISBN 978-1-83968-832-4
eBook (PDF) ISBN 978-1-83968-833-1

For EU product safety concerns:
IN TECH d.o.o., Prolaz Marije Krucifikse Kozulić 3, 51000 Rijeka, Croatia, info@intechopen.com or visit our website at intechopen.com.

We are IntechOpen, the world's leading publisher of Open Access books

Built by scientists, for scientists

5,600+
Open access books available

137,000+
International authors and editors

170M+
Downloads

156
Countries delivered to

Our authors are among the
Top 1%
most cited scientists

12.2%
Contributors from top 500 universities

WEB OF SCIENCE™

Selection of our books indexed in the Book Citation Index (BKCI)
in Web of Science Core Collection™

Interested in publishing with us?
Contact book.department@intechopen.com

Numbers displayed above are based on latest data collected.
For more information visit www.intechopen.com

Meet the editors

Dr. Hu Huang received his BE and Ph.D. in Mechanical Engineering from Jilin University, China, in 2010 and 2014, respectively. Supported by the Japan Society for the Promotion of Science (JSPS), he worked as a research fellow at Keio University, Japan, from 2014 to 2017. Currently, he is a full professor at Jilin University. His research interests include precision instruments, mechanical properties of materials (metallic glass, hard-brittle materials, etc.), and surface modification of materials. He has published more than eighty peer-reviewed journal papers indexed by the Science Citation Index Expanded (Web of Science). He is a senior member of the Chinese Mechanical Engineering Society, and a member of the International Society of Bionic Engineering, the Institute of Electrical and Electronics Engineers (IEEE), and the Japan Society for Precision Engineering.

Jianping Li graduated with a Ph.D. in Engineering from Jilin University, China, in 2016. Since then, he has obtained membership in the Japan Society for the Promotion of Science (JSPS) and served as a special researcher for foreigners at the Japanese National Chiba University. In 2018, he attended the College of Engineering, Zhejiang Normal University, where he is now an associate professor. He has been engaged in the research of micro/nano-sensing and driving, bioelectrical impedance measurement, and medical application technologies, especially piezoelectric actuators. To this date, he has published more than seventy academic papers and authorized one US patent and fourteen Chinese patents.

Contents

Preface

Precision positioning is widely required in both scientific research and industrial application. Taking advantage of high resolution, rapid response, and compact structure, piezoelectric actuators are broadly employed for achieving precision positioning, for example, in the fields of precision machining and measurement, nanomechanical testing, atomic force microscopy (AFM), micromanipulation, and so on. To satisfy the increasing requirements in precision positioning, great efforts have been made to improve the performances of piezoelectric actuators. For example, more and more novel principles have been proposed to design piezoelectric actuators. With the development in the piezoelectric material, driving principle, structure design, and control strategy, the speed, positioning resolution/accuracy, loading capacity, working bandwidth, and stroke of piezoelectric actuators have been significantly improved, further enhancing their applications. Over eight chapters, this book discusses recent achievements and developments in the field.

Chapter 1 focuses on tuning the magnetoelectric coefficient in sintered piezoelectric–magnetostrictive composites by introducing some individual phases. It examines the effects of these introduced phases on the magnetoelectric coefficient, coercive field, saturation magnetization, magnetic permeability, piezomagnetic coefficient, and more.

Chapter 2 focuses on the parasitic motion principle (PMP) of piezoelectric actuators. This type of piezoelectric actuator has received more attention in recent years because of its simple structure, control, and flexible design. The chapter defines PMP as well as discusses its application in the design of piezoelectric actuators. It also presents the similarities and differences of PMP with two other types of stepping principles. Finally, the chapter considers recent developments in structural design and performance improvements of PMP piezoelectric actuators and points out existing issues and future research directions.

Chapter 3 discusses inchworm-type piezoelectric actuators, examining their motion principles, classification (linear, rotary, and multi-DOF), and development. It also examines the future directions of these actuators.

Chapter 4 focuses on the stick-slip piezoelectric actuator, which is very promising for achieving both long working stroke and high positioning resolution. In this chapter, the authors summarize their recent studies on this topic both on the structural design and driving method. In terms of structural design, the authors present various flexure hinge mechanisms, especially asymmetrical flexure hinges, to improve the performances of stick-slip piezoelectric actuators, for example, their velocity and loading capacity. In terms of driving methods, they also present a non-resonant mode smooth driving method (SDM) to suppress the backward motion, and a resonant mode SDM to improve the output performances. The authors present many solutions for improving the performances of stick-slip piezoelectric actuators.

Chapter 5 models piezoceramic actuators for potential applications in future control. It develops a full electro-mechanical model of piezoceramic actuators based on previous studies and determines the parameters of this model. The output of the model agrees well with the experimental data, verifying the validity of the proposed model.

Chapter 6 proposes a coupled suspension system and piezoelectric model to predict the potential of harvested electric power in vehicle suspension systems. It investigates the performance of the piezoelectric material PZT-5H in respect to harvesting energy based on energy density and electric power density.

Chapter 7 presents the applications of piezoelectric actuators in microfluidic technology. It discusses some structures and operating principles of single-chamber and multi-chamber piezoelectric micropumps and demonstrates their application in a chip water-cooling system.

Chapter 8 focuses on the roles of piezoelectric ultrasonic motors (USMs) in the Industry 4.0 era. It discusses their novel working principles, illustrates examples for their effective utilization, and analyzes the key Industry 4.0 technologies for improving their performance.

I am very grateful for the help of several colleagues who contributed to this book, which we hope will assist students and researchers in their work, as well as contribute to new developments in the field of piezoelectric actuators.

Hu Huang
School of Mechanical and Aerospace Engineering,
Jilin University,
Changchun, China

Jianping Li
College of Engineering,
Zhejiang Normal University,
Jinhua, China

Section 1

Piezoelectric Materials

Chapter 1

Doping Effect on Piezoelectric, Magnetic and Magnetoelectric Properties of Perovskite—Ferromagnetic Magnetoelectric Composites

Rashed A. Islam

Abstract

This chapter explains the effect of compositional modification on the magnetoelectric coefficient in sintered piezoelectric – magnetostrictive composites. It was found that 15 at% doping of $Pb(Zn_{1/3}Nb_{2/3})O_3$ [PZN] in $Pb(Zr_{0.52}Ti_{0.48})O_3$ [PZT] enhances the piezoelectric and magnetoelectric properties of a PZT – 20 at% $Ni_{0.8}Zn_{0.2}Fe_2O_4$ [NZF] composite. The effect of doping on the ferromagnetic phase was also investigated. With increases in Zn concentration, it was found that the coercive field and Curie temperature of $Ni_{(1-x)}Zn_xFe_2O_4$ [NZF] decreases, while its saturation magnetization has a maxima at 30 mole% Zn. X-ray diffraction revealed that the lattice constant of NZF increases from 8.32 Å for 0 at% Zn to 8.39 Å for 50 at% Zn. The magnetoelectric coefficient was found to have a maxima of 144 mV/cm.Oe at 30 at% Zn. To understand better, the effect of 40% (by mole) Zn substitution on structural, piezoelectric, ferromagnetic and magnetoelectric properties of Pb $(Zr_{0.52}Ti_{0.48})O_3$ - $CoFe_2O_4$ (PZT - CFO) sintered composite is also explained. X-ray diffraction of $Co_{0.6}Zn_{0.4}Fe_2O_4$ (CZF) showed the shift in almost all diffraction peaks to lower diffraction angle confirming the increase in lattice parameter in all three direction from 8.378 (for CFO) to 8.395 Å for $(Co,Zn)Fe_2O_4$ (CZF). SEM and TEM results showed defect structure (cleavage, twins, strain fields) in the CZF particle, which is a clear indication of misfit strain developed due to lattice expansion. Magnetic properties measured over temperature (5 K – 1000 K) showed increased magnetization but lower magnetic Curie temperature in PZT - CZF particle. Magnetoelectric coefficient measured as function of ferrite concentration showed an increase of more than 100% after doping the CFO phase with 40% Zn. This enhancement can be attributed to increase in the lattice strain, magnetic permeability and decrease in coercivity.

Keywords: piezoelectric constant, dielectric constant, ferromagnetic, magnetoelectric coefficient, misfit strain

1. Introduction

Magnetoelectric [ME] particulate composites combine the magnetostrictive and the piezoelectric properties of materials, through product tensor properties [1].

IntechOpen

Multiferroic magnetoelectric materials possess two or more ferroic properties such as ferroelectricity, ferromagnetism and ferroelasticity [2–4]. The spin lattice structure in a magnetoelectric composite can be directly related to (i) linear or non-linear shape change in magnetostrictive phase under alternative magnetic field, (ii) polarization change in piezoelectric phase through field induced alternating stress–strain and finally (iii) charge developed in the piezoelectric phase due to this alternating stress [1, 5, 6]. The interrelationship between ferroelectricity and magnetism allows magnetic control of ferroelectric properties and vice-versa. Single phase magnetoelectrics such as Cr_2O_3, $BiFeO_3$, $YMnO_3$ etc. exhibit poor combination of electric and magnetic properties at room temperature [7–9]. On the other hand, two-phase magnetoelectric (ME) materials provide large coupling and may play important role in future magnetoelectric devices [10]. Another important issue that can be very influential not only in nanostructures or multilayer structure but also in bulk ceramic composite is the interface chemistry. Migration of mobile atoms (from ferroelectric and magnetic phases) through the interface causes ferroelectric and magnetic instability and alters the interface chemistry, which affects the interface magnetoelectric properties [10, 11]. There are lot of advantages that sintered particulate offers, compared to in-situ composites (i.e. unidirectionally solidified of $BaTiO_3 - CoFe_2O_4$), such as they are cost effective to produce, fabrication is easy and finally and most importantly the process parameters can be controlled much better. In terms of ME responses, laminate magnetoelectric composites gained a lot of popularity and can be fabricated by attaching piezoelectric layer between two layers of magnetostrictive discs or plates. Sintered particulate composites exhibits low resistivity, defects, diffusions at the interface and incompatibility of elastic compliances and mismatch in coefficient of thermal expansion. As a result, sintered composites show inferior ME responses compared to laminated composites. Therefore, it is essential to augment the composition, grain size, grain orientation, and sintering conditions in order to enhance the Magnetoelectric properties of the sintered composites.

The composites exploit the product property of the materials [12–14] where the ME effect can be realized by mixing individual piezomagnetic and piezoelectric phases or individual magnetostrictive and piezoelectric phases. In early 70s, researchers at Philips Laboratories demonstrated ME composites [15–18] by unidirectional solidification of eutectic composition of $BaTiO_3 - CoFe_2O_4$. The results showed a high ME voltage coefficient dE/dH of 50 mV/cm•Oe with 1.5 wt % of excess of TiO_2 [15]. Later an even higher ME coefficient of 130 mV/cm•Oe was obtained in eutectic composition of $BaTiO_3$-$CoFe_2O_4$ by unidirectional solidification [17]. Currently, various particulate composites consisting of piezoelectric and magnetostrictive materials with different connectivity schemes including "3-0" and "2-0" have been reported, using $LiFe_5O_8$, $NiFe_2O_4$, $(Ni,Zn)Fe_2O_4$, $CoFe_2O_4$, $CuFe_2O_4$ as magnetostrictive materials and $BaTiO_3$, $Pb(Zr,Ti)O_3$ as piezoelectric phase [15–24].

The figure of merit for a ferromagnetic-ferroelectric composite is large magnetoelectric coefficient (i.e., susceptibility) given as:

$$\text{Figure or merit} = \sqrt{\mu\varepsilon}; \tag{1}$$

Here μ = ferromagnetic permeability and ε is the dielectric permittivity. It is eminent from Eq. (1) that a high dielectric constant piezoelectric phase and a high permeability magnetic phase would produce a composite with optimum ME response if we can keep the high resistivity, low interface defects and lower rate of interface diffusion. Literatures and experimental review showed that the Nickel

Ferrites are stable in PZT up to 1250°C and offers higher permeabilities. Also in terms of resistivity and loss, Ni based ferrite are preferable over Mn based ferrite, because Ni based Ferrites have higher electric resistivity and lower dielectric loss.

2. Principles of magnetoelectricity

The thermodynamic consideration of magnetoelectric effect is obtained from the expansion of free energy of the system in terms of magnetic and electric field, such as.

$$G\left(\vec{E},\vec{H}\right) = G_o - P_s\_E - M_s\_H - \frac{1}{2}\chi_{ij}(E)\_E_i\_E_j - \frac{1}{2}\chi_{ij}(H)\_H_i\_H_j - \alpha_{ij}\_E_i\_H_j \quad (2)$$

where and are the electric field and magnetic field respectively. Differentiation of Eq. (2) gives us polarization and magnetization as following:

$$P_i = -\left(\frac{\delta G}{\delta E}\right) = P_s + \chi_{ij}E_j + \alpha_{ij}H_j \quad (3)$$

$$M_i = -\left(\frac{\delta G}{\delta H}\right) = M_s + \chi_{ij}H_j + \alpha_{ij}E_j \quad (4)$$

Here α_{ij} is the magnetoelectric tensor. Magnetoelectric effect combines two important materials property, permittivity and permeability, and for a single-phase material they define the upper limit of α_{ij} as following.

$$\alpha_{ij} < \sqrt{\varepsilon_{ij}\mu_{ij}} \quad (5)$$

Single phase multiferroic materials shows either low permeability or low permittivity or both. As a result, the magnetoelectric coupling is small. For high response dual phase magnetoelectric materials, combination between ferroelectric and ferromagnetic phase need to be established via strain. Piezoelectric coefficient (d_{33}/d_{31}) defines the materials property that converts applied stress in to proportional electric charge. The linear equations for a piezoelectricity and magnetostrictions are given as:

$$D_3 = \varepsilon_{33}^T E_3 + d_{33}T_3 \quad (6)$$

$$S_3 = d_{33}E_3 + s_{33}^E T_3 \quad (7)$$

$$S = s^H T + qH \quad (8)$$

$$B = qT + \mu^T H \quad (9)$$

where
D = the dielectric displacement,
E = electric field,
T = stress,
S = strain,
ε = permittivity,
s = elastic compliance,
d = is piezoelectric charge constant.

B = magnetic induction.
q = piezo magnetic coefficient.
μ = permeability and.
H = magnetic field.

Magnetoelectric coefficient of a composite can be described in direct notation of tensors as:

$$T = cS - e^T E - \alpha S^{ms} \tag{10}$$

$$D = eS + \varepsilon E + \alpha H \tag{11}$$

$$B = \mu(\varepsilon, E, H)H \tag{12}$$

where σ, c and K are the stress, stiffness constant at constant field and dielectric constant at constant strain respectively. It was found in the literatures that ME coefficient can be varied by piezoelectric and piezomagnetic coefficients. Elastic compliances (s) of piezoelectric and magnetostrictive phases are found to be another critical parameter that affects the ME coefficient. According to Srinivasan et al., the ME coefficient can be written as:

$$\frac{\delta E_3}{\delta H_1} = \frac{-2d_{31}^{p} q_{11}^{m} v^{m}}{\left(s_{11}^{m} + s_{12}^{m}\right)\varepsilon_{33}^{T,P} v^{p} + \left(s_{11}^{p} + s_{12}^{p}\right)\varepsilon_{33}^{T,P} v^{m} - 2\left(d_{31}^{p}\right)^{2} v_{m}} \tag{13}$$

where
d_{31}^{p} = piezoelectric coefficient,
v^m and v^p = volume of magnetic and piezoelectric phase,
t^m and t^p = thickness of magnetic and piezoelectric phase,
s_{11}^{p}, s_{12}^{p} = the elastic compliances for piezoelectric phase,
s_{11}^{m}, s_{12}^{m} = elastic compliances for magnetostrictive phase,
q_{11} = piezomagnetic coefficient of the magnetic phase and.
$\varepsilon_{33}^{T,P}$ = permittivity of the piezoelectric phase.

Further derivation for the magnetoelectric coefficient in T – T mode of Eq. (13) was done by Dong et al. and was expressed as:

$$\left|\frac{dV}{dH}\right|_{T-T} = \beta \frac{n(1-n)Ad_{33,m}d_{31,p}g_{31,p}}{S_{11}^{E}\left[nS_{11}^{E}\left(1-k_{31}^{2}\right) + (1-n)S_{11}^{H}\right]} \tag{14}$$

where
β = a constant related to DC magnetic field (<1),
n = the ratio of magnetostrictive layer thickness to the composite thickness,
d = the piezoelectric strain constant,
s = the elastic constant,
g = the piezoelectric voltage constant,
A = the cross-sectional area of the laminate and.
k = is the electromechanical coupling factor.

It is quite clear from the Eq. 14 that the ME coefficient is directly related to piezoelectric constant (d_{31}) and piezomagnetic coefficient (q_{11}). d_{31}^{p} which is related to dielectric permittivity $[d_{31}^{2} = k_{31}^{2}(s_{11}^{E}\varepsilon_{33}^{T})]$ and q_{11} is related to permeability $[q_{11,m} = \mu_{33} \cdot s_{33} \cdot \lambda_{33}]$.

3. Effect of doping

As is well known that compared to $BaTiO_3$, PZT has stronger piezoelectric and dielectric properties, higher Curie temperature, higher resistivity and lower sintering temperature. Doping in PZT can be done by adding acceptor dopants (Fe, Mn, Ni, Co) or donor dopants (La, Sb, Bi, W, Nb) in order to make it piezo-electrically hard or a soft. Hard piezoelectric materials can be characterized as decreased dielectric constant and loss, lower elastic compliance, lower electromechanical coupling factor, and lower electromechanical losses compared to undoped PZT. Soft piezoelectric materials exhibit increased dielectric constant, dielectric loss, elastic compliance, electromechanical coupling factor, and electromechanical losses. **Table 1** shows a comparison chart how the physical, dielectric and piezo-electric properties vary between soft and hard piezoelectric materials.

The open circuit output voltage (V), under an applied force of a ceramic is given as:

$$V = E \cdot t = -g \cdot X \cdot t = -\frac{g \cdot F \cdot t}{A} \tag{15}$$

where
t = the thickness of the ceramic,
E = the electric field, and.
g = the piezoelectric voltage coefficient given as:

$$g = \frac{d}{\varepsilon_o \varepsilon^X} \tag{16}$$

where ε^X is the dielectric constant under constant stress condition.
The charge (Q) generated on the piezoelectric ceramic is given by the relation:

$$Q/_V = \frac{\varepsilon^X \varepsilon_o A}{t} = C \tag{17}$$

where

$$\text{C} = \text{the capacitance and} \tag{18}$$

Properties	Soft Piezoelectric	Hard Piezoelectric
Electrical Resistance	Higher	Lower
Permittivity	Higher	Lower
Dielectric Constants	Superior	Inferior
Dielectric Loss	Higher	Lower
Piezoelectric Constants	Superior	Inferior
Coercive Field	Lower	Higher
Mechanical Quality Factor	Lower	Higher
Electromechanical Coupling Factor	Larger	Smaller
Linearity	Poor	Better

Table 1.
Comparison of Dielectric and Piezoelectric properties between Soft and Hard Piezoelectric Materials.

It can be inferred from Eq. (17) that a piezoelectric plate can behave like a parallel plate capacitor at low frequencies. From here it can be derived that under ac stress, electric energy generated is given as:

$$U = \frac{1}{2}CV^2$$

or energy per unit volume,

$$u = \frac{1}{2}(d.g) \cdot \left(\frac{F}{A}\right)^2 \tag{19}$$

Eq. (15) and (19) conclude that under a fixed AC mechanical stress, piezoelectric material with high (d.g) product and high piezoelectric voltage (g) constant will generate high voltage and high power for a fixed area and thickness. In the case of magnetoelectric composite, the force is applied on the piezoelectric phase due to magnetostriction through elastic coupling, therefore the high energy density piezoelectric material will lead to higher response.

Nickel and cobalt ferrites have the advantage of higher resistivity and increased permeability. Cobalt ferrite has higher magnetization but also has higher coercivity compared to nickel ferrite. In order to increase the resistivity, permeability, and magnetization, doping of zinc in to ferrite is beneficial but it also reduces its magnetic Curie temperature. The theory behind this is, Zn^{+2} replaces Fe^{+3} on the tetrahedral sites as it is added to the spinel structure and Fe^{+3} occupies the vacant octahedral sites emptied by Co^{+2}. As a result, there will be no unpaired electrons for Zn^{+2}, Co^{+2} has one and Fe^{+3} has five. Hence the outcome of it is increase in magnetization of Zn-doped ferrites.

4. Synthesis and fabrication

$Pb(Zr_{0.52}Ti_{0.48})O_3$ (PZT), $0.85[Pb(Zr_{0.52}Ti_{0.48})O_3] - 0.15[Pb(Zn_{1/3}Nb_{2/3})O_3$ [PZT (soft)], $Pb(Zr_{0.56}Ti_{0.44})O_3 - 0.1\ Pb[(Zn_{0.8/3}\ Ni_{0.2/3})Nb_{2/3}]O_3$ + 2 (mol %) MnO_2 [PZT (hard)], $Ni_{(1-x)}Zn_xFe_2O_4$ [NZF] (where x varies from 0 to 0.5) and $Co_{(1-y)}Zn_yFe_2O_4$ [CZF], were synthesized using mixed oxide route. PZT, NZF and CZF powders were calcined at 750°C for 2 hrs and 1000°C for 5 hrs, respectively in order to make sure that inorganic oxides react to each other. Powder X-ray diffractions patterns were taken using Siemens Krystalloflex 810 D500 diffractometer to make sure that the pure perovskite and pure spinel structure was formed out of PZT/PZT (soft)/PZT (hard) and NZF/CFO/CZF. The PZT and NZF powders were then mixed together as 0.8 PZT – 0.2 NZF and compacted. CFO/CZF powders were mixed in PZT with stoichiometric ratio of 3, 5, 10, 15 and 20 mole percents. After homogeneous mixing using ball mill, powder was pressed using a hardened steel die having diameter of 12.7 mm under a pressure of 2 ksi and then cold isostatically pressed under pressure of 40 ksi. This was followed by pressure-less sintering in air at 1150°C for 2 hrs, resulting in consolidated ceramic composites. XRD patterns of sintered samples showed only two phases (PZT and CFO/CZF). Ag/Pd paste was painted manually on top and bottom of the sintered disc using a paint brush and heated for an hour at 825° C. The polarization process was done in a heated (120°C) silicone oil bath. D.C. electrical field of 2.5 kV/mm for 20 minutes was applied for the poling process. Dielectric constant as a function of temperature was measured using HP 4274A LCR meter (Hewlett Packard Co. USA). Magnetization as a function of temperature was measured using Quantum Design physical properties measurement system from room temperature to 900 K. Transmission electron microscopy (TEM) was conducted by using JEOL 1200EX machine with an accelerated voltage of 120 kV.

5. Results and discussions

Figure 1(a) [25] Compares the polarization of undoped, hard and soft PZT. The max polarizations observed were 14.61, 23.54 and 31.65 μC/cm^2 respectively for these three groups. It was also recorded that elastic compliance (S_{11}) of these three compositions are 1.74 x 10^{-11} (soft), 1.37 x 10^{-11} (hard) and 1.11 x 10^{-11} (undoped) m^2/N respectively. Due to the presence of metal vacancies, soft PZT's have higher polarizations per unit field applied, resulting in enhancement in both the dielectric and the piezoelectric susceptibilities.

An increase in piezoelectric constant (d_{33}) (from 75 to 105 pC/N) and increase in dielectric constant (from 642 to 914) was observed when an undoped PZT was doped with PZN to make it soft. These increase in dielectric and piezoelectric properties clearly reflected in Magnetoelectric coefficient of PZT (soft) – 20 NZF composite $\frac{\delta E}{\delta H}$ = 186.5 mV/cm.Oe. as observed in **Figure 1(b)**. Due to increase in electromechanical coupling, smaller coercive field, and superior dielectric and piezoelectric properties Soft PZTs show larger ME voltage coefficient compared to hard (154 mV/cm.Oe) or undoped PZT (128 mV/cm.Oe).

Microstructure of PZT (undoped) – 20NZF and PZT (soft) – 20NZF composites were dense with the measured densities of ≥95% [25]. And the average grain size of the PZT – 20NZF composite was about 800 - 850 nm, whereas that of PZT (soft) – 20 NZF was smaller and the average ranges between 650 and 700 nm. Doping of PZN in PZT reduces the grain size as both the composite were sintered at 1150°C for 2 hours. In one of our previous studies it was shown that above 600 nm, the ME coefficient does not change much with increase in grain size [26]. So, both PZT (undoped) – 20NZF and PZT (soft) – 20NZF composites have optimum grainsize in terms of ME coefficient. Besides the grainsize reduction, the resistivity of PZT increases upon doping with PZN, as lower leakage currents were observed after poling (**Figure 2**).

Figure 3 [25] shows the saturation magnetization (M_s), coercive field (H_c) and magnetic Curie Temperature of the PZT – 20NZF composite as a function of Zn doping in NZF. It is clearly observed that the coercive field starts to drop as we increase the Zn doping in NZF. On the other hand, saturation magnetization becomes optimum (0.72 emu/gm) at around 30% Zn doping in NZF and then starts

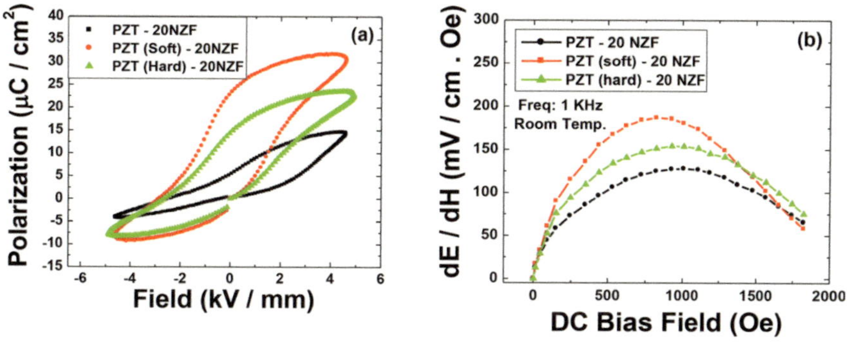

Figure 1.
(a) Polarization vs. electric field loop and (b) ME coefficient vs. DC bias of different compositions of PZT – 20 NZF composites [25].

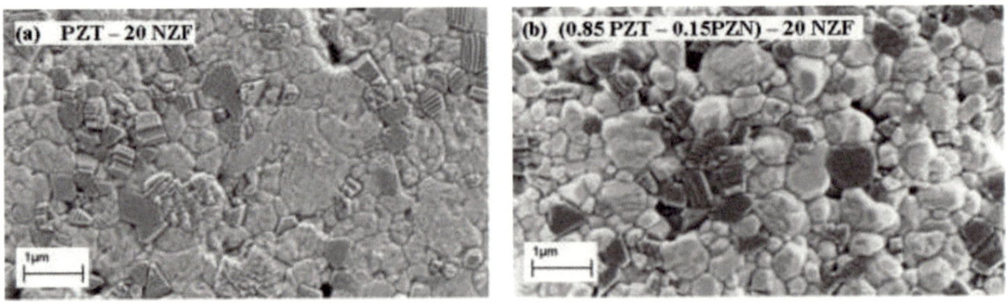

Figure 2.
Microstructure of (a) PZT – 20 NZF and (b) (0.85PZT –0.15 PZN) – 20 NZF [25].

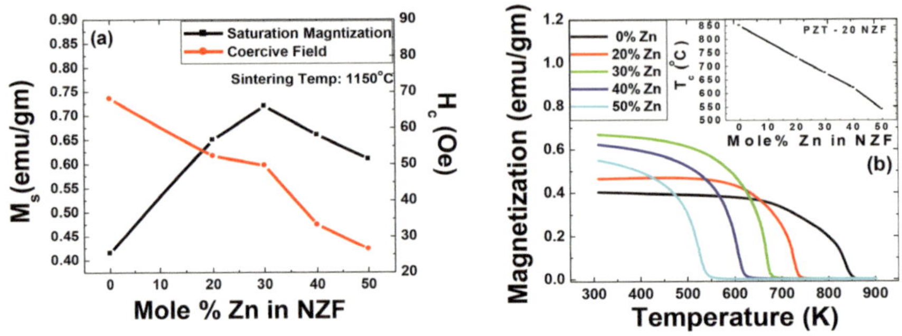

Figure 3.
Magnetic properties as a function of Zn doping. (a) M_s and H_c vs. Zn concentration and (b) magnetization vs. temperature [25].

to drop off with increase in Zn doping. In terms of Ferromagnetic Curie Temperature, it started to drop from 850 K to 549 K as the Zn concentration was increased from 0 to 50 mole %.

Figure 4(a) [25] Shows the hkl = (400) diffraction peak for NZF composites as a function of Zn concentration. The (400) peak started to shift towards lower Bragg angles as Zn concentration was increased which indicated an enlargement of lattice parameters. From Bragg's law, we determine a 0.9% lattice expansion (8.32 Å for $NiFe_2O_4$ and 8.394 Å for $Ni_{0.5}Zn_{0.5}Fe_2O_4$) with this change in crystal chemistry. **Figure 4(b)** shows the ME voltage coefficient as a function Zn concentration in PZT – 20 NZF. 30 at% Zn concentration in NZF showed the maximum value of $\frac{\delta E}{\delta H}$ = 138 mV/cm.Oe, whereas 0 at% Zn concentration in NZF (pure Ni ferrite) has a

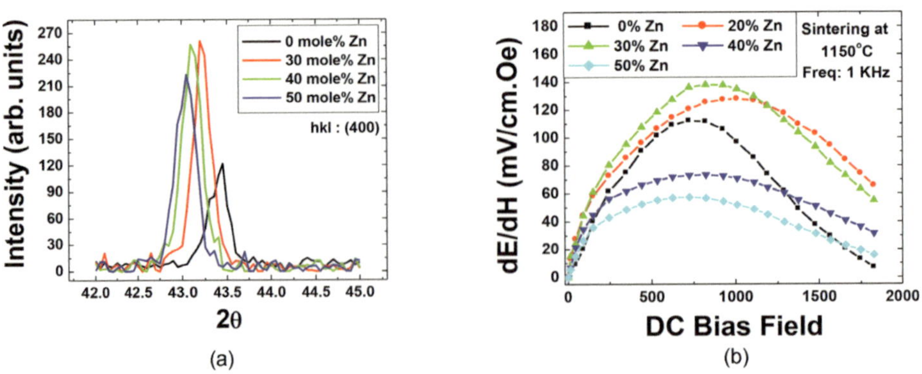

Figure 4.
Effect of Zn concentration on (a) peak shift of 400 peaks and (b) magnetoelectric coefficient of PZT – 20 NZF [25].

maximum value of 112 mV/cm.Oe. With Zn concentration above 30%, the ME coefficient dropped notably, reaching a value of 60 mV/cm.Oe for 50 at% Zn. Saturation magnetization and magnetoelectric responses as a function of Zn concentration in NZF was found to have similar effect because permeability μ is directly related to magnetization M, via $\mu = 1 + 4\pi\frac{M}{H}$. Furthermore, the changes in the ME coefficient with Zn concentration depends on the change in effective piezomagnetic coefficient ($d_{33,m} = \mu_{33}.s_{33}.\lambda_{33}$) which is directly related to the permeability.

Figure 5 (a-c)show the density, dielectric constant and piezoelectric constant as a function of mole percent ferrite in the composite for two different compositions. It is clear that as the ferrite concentration increases, density, dielectric constant and piezoelectric constant decreases. All the compositions showed more than 98% of the theoretical density and the microstructural analysis confirmed this measurement. For PZT – CFO, there is a slight increase in density from 3–5% concentration which can be attributed to better sintering as the.

ferrite becomes more homogenized in the matrix. The density increase was also observed when the composition changed from 10–15% for CFO in PZT – CFO composite. This can be explained by grain coarsening of PZT. As the CFO content increases from 10–15% there is slight increase in dielectric constant and then with further increase in ferrite concentration, dielectric constant starts to drop. It is well-known that grain coarsening has direct effect on dielectric properties, whereby, dielectric constant increases with larger grain size. There was no significant difference in piezoelectric data between CFO and CZF based ferrite composites with ferrite concentration.

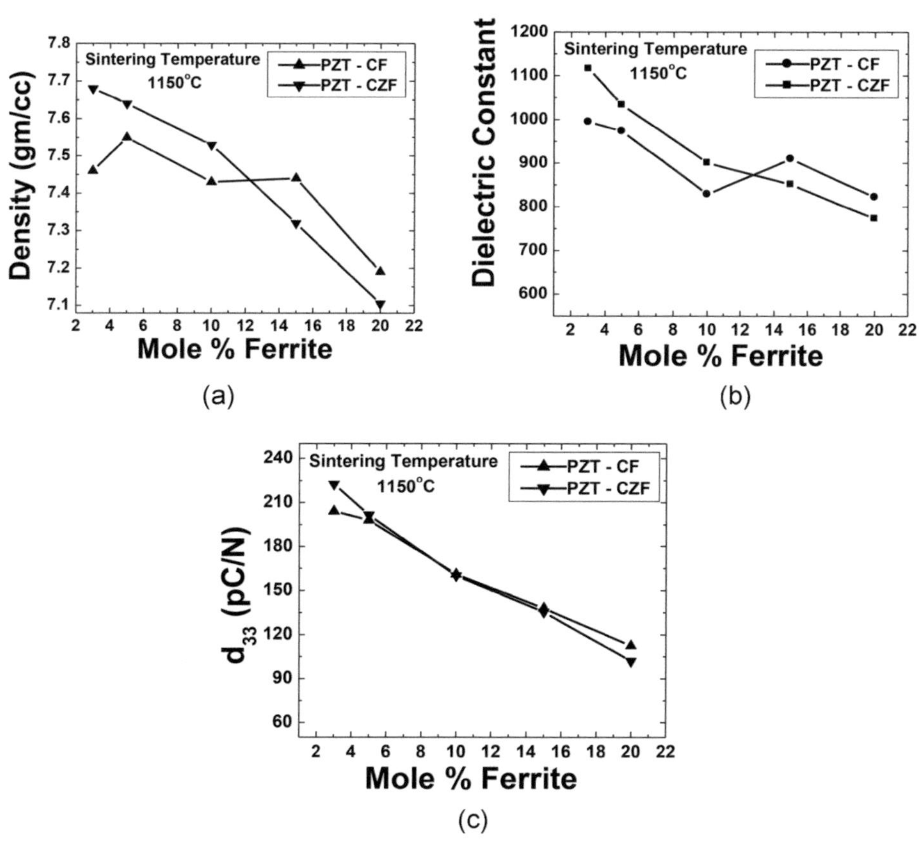

Figure 5.
(a), (b) and (c) The density, dielectric constant and piezoelectric constant as a function of mole percent ferrite.

A comparison between PZT – CFO and PZT – CZF in terms of room temperature magnetic properties is presented in **Figure 6(a)** and **(b)** for 3% and 5% ferrite concentration. In both the cases, a considerable difference in coercivity between CFO and CZF particles was observed. For 3% ferrite concentration the saturation magnetization of CFO was slightly higher than CZF but the coercivity was much lower (33 Oe compared to 263 Oe). For 5% CFO and CZF concentration, coercive fields of 53 and 288 Oe were measured and the saturation magnetization of CFO particle (76.6 memu) was slightly lower than the CZF particle (88.54 memu). **Figure 6(c)** and **(d)** show the magnetization of PZT – 5 CFO and PZT – 5 CZF composites from 5 K to 300 K and from 310 K to 1000 K respectively. It is quite interesting to observe that the magnetization for both the composites start to drop from 5 K to 300 K. The drop for PZT – 5CFO is linear - 0.04 emu to 0.035 emu and for PZT – 5CZF is non –linear 0.075 to 0.065 emu respectively for 70 mg of sample weight. There is a slope change in the PZT – CZF magnetization curve at around 150 K – which is close to the curie temperature of $ZnFe_2O_4$. From 150 K to 5 K – an increased slope was observed. Below the curie temperature, Zn ferrite also contributed to the magnetization curve. Besides the increase in magnetization in subzero temperature for both the composites can also be explained by the atomic vibration of the crystal lattice. As the temperature starts to drop below the room temperature and approaches the absolute zero temperature, the atomic vibration is seized, resulting in much stable crystal lattice, which gives us accurate measurement of the magnetization. At high temperature (from 310 to 1000 K) the magnetization of PZT – CFO starts to increase and then decrease to zero at Curie temperature, which is almost 750 K. The PZT – CZF Curie temperature was recorded at 450 K. This drop

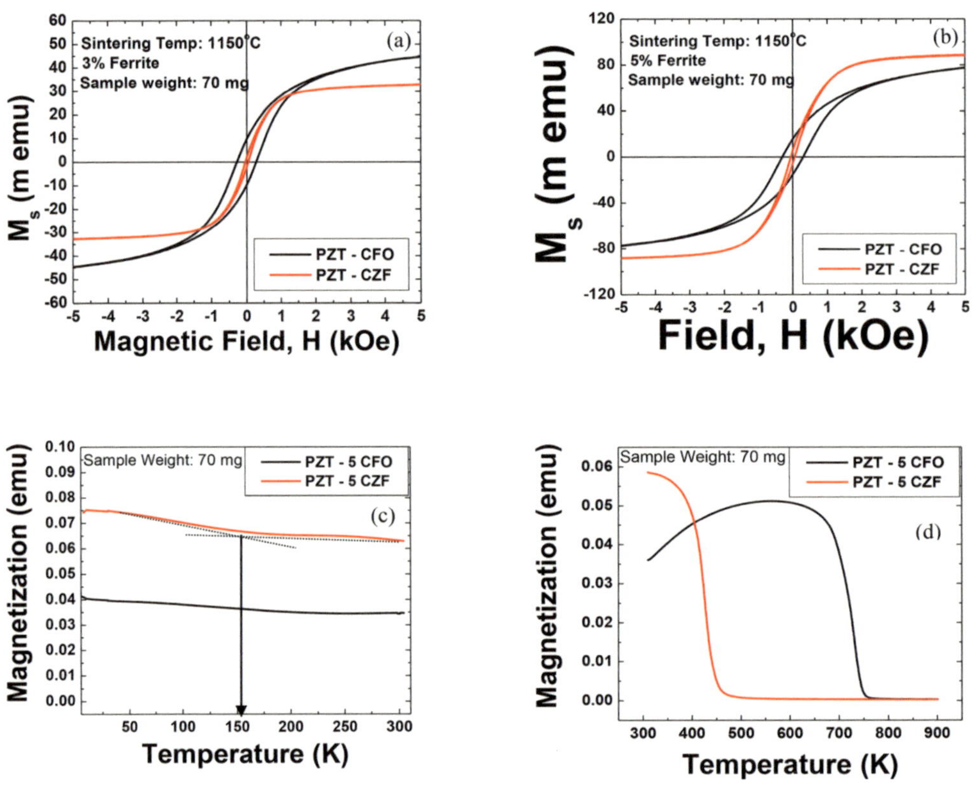

Figure 6.
(a) and (b). Hysteresis loop for PZT – CFO and PZT – CZF. (c) and (d): the magnetization of PZT – 5 CFO and PZT – 5 CZF composites.

in Curie temperature from 750 to 450 K is due to the substitution of Zn on the cobalt site. The substitution of Zn^{2+} for Fe^{3+} reduces the Curie temperature of the ferrite [6, 7, 9]. On the other hand, increasing the zinc content of cobalt-zinc ferrites increases their lattice parameter while decreasing the saturation magnetization above 50% Zn due to augmented B-B interaction followed by reduced A-B interaction. Also, the presence of Co^{2+} ion in the cobalt-zinc ferrite hastens the $Co^{2+} + Fe^{3+} \Leftrightarrow Co^{3+} + Fe^{2+}$ exchange reaction in octahedral sites, while tetrahedral sites are preferentially occupied by zinc cations. Tetrahedral sites in the spinel structure are suitable for cationic radii in the range of 0.58 Å to 0.67 Å, while octahedral sites can accept cations with radii in the range of 0.70 Å to 0.75 Å [22]. Therefore, in the unit cell structure, Co^{2+} (0.72 Å) and Fe^{2+}(0.75 Å) may replace Zn^{2+}(0.74 Å), while Co^{3+}(0.63 Å) can exchange sites with Fe^{3+}(0.64 Å). This exchange in Co-Zn ferrite system, the substitution of non-magnetic zinc in place of ferromagnetic cobalt leads to a decrease in Curie temperature owing to diminishing A-B super exchange interaction.

Figure 7 shows the XRD pattern for CFO and CZF particles. Inset shows magnified 311 peaks for CFO and CZF particles respectively. The shift in peaks to lower angle for CZF particles is clearly noticeable which results in a larger unit cell size for CZF particles. The increase in unit cell can induce strain. Strain can be revealed as strain fields or cleavage or other defects inside the microstructure. **Figure 8** shows the ME coefficient as a function.

of ferrite concentration. For PZT – CFO the maximum ME coefficient of 25 mV/cm.Oe was recorded at 15% ferrite concentration, which drops again for 20% ferrite. The measured ME coefficient is quite low compared to the PZT – Nickel ferrite composites. Cobalt ferrite has a very high coercive field compared to the nickel ferrite composition. Thus, a high DC bias field is necessary to obtain the peak ME coefficient. Another contributing factor is the initial permeability. Cobalt ferrite has lower initial permeability than the nickel ferrite which contributes towards lower ME coefficient.

Figure 9(a) and **(b)** show the SEM images of the microstructure of the PZT-20CFO and PZT – 20CZF samples sintered at 1125°C respectively. Dense microstructures for both compositions were obtained and the sintered samples had grain size of 1 to 1.5 μm. Elemental analysis using the EDX showed that CFO and CZF

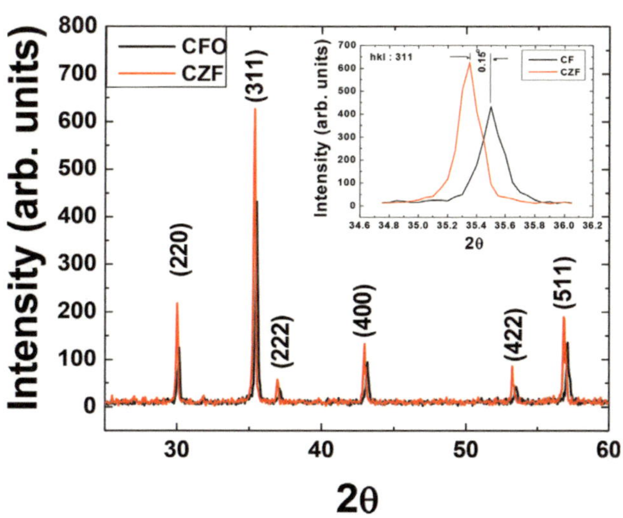

Figure 7.
XRD patterns for CFO and CZF particles.

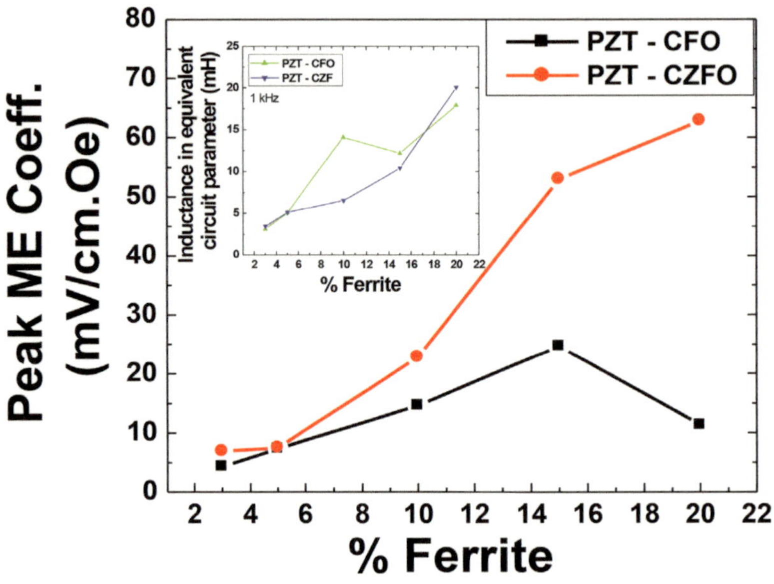

Figure 8.
ME coefficient as a function of ferrite concentration.

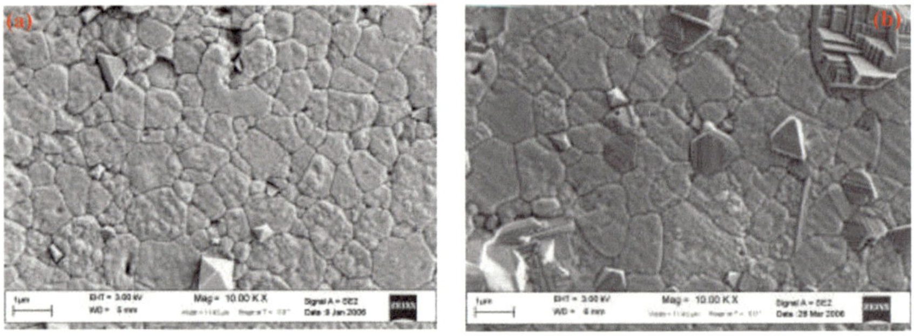

Figure 9.
(a) and (b): SEM images of the microstructure of the PZT-20CFO and PZT – 20CZF.

grains are distributed in the piezoelectric matrix. **Figure 10(a)** and **(b)** shows the bright field TEM images of the sintered.

PZT – 20 CFO and PZT – 20 CZF samples respectively. Compared to PZT – CFO, PZT - CZF sintered samples were found to consist of twin boundaries, cleavage, and strain fields at the interface of PZT and CZF grains. These defects develop to accommodate the mismatch in the PZT and CZF lattices, as ferrite (CFO/CZF) lattice parameters are more than double the lattice parameter of the PZT lattice. The lattice parameter increases by 0.2% for 40% Zn doped sample as observed in the XRD patterns. In the inset, diffraction pattern of a PZT grain is shown. No superlattice diffraction spots were observed near the first order diffraction spots, which indicate less intense diffusion level. From the SAED diffraction pattern the lattice parameter a and c are calculated as 4.05 and 4.132 Å, hence the c/a ratio is 1.02. Larger width domain patterns were also observed near the interface, which is characteristic of 90° domains. Besides that, intergranular heterogeneity in domain width is observed all over the structure especially near the interface. The observed defects in PZT - 5% CZF are in line with the SEM images. A finer scale domain structure, which usually has striation like morphology and periodically spaced was

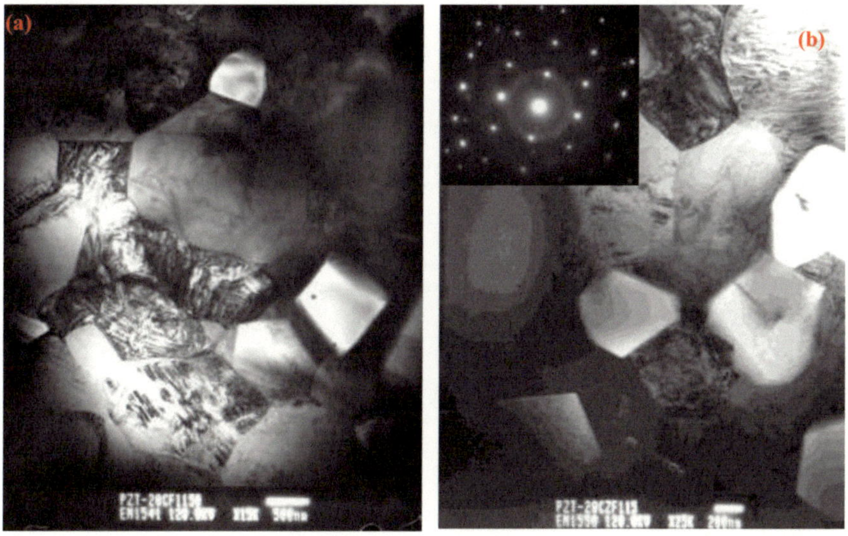

Figure 10.
(a) and (b): bright field TEM images of the sintered PZT – 20 CFO and PZT – 20 CZF.

observed in this structure away from the interface. These finer domains appear when the stress is relieved from the structure. It can be inferred that near the ferroelectric – ferromagnetic interface, the stress is higher and defects are observed due to strain mismatch, whereas the area away from the interface has lower stress.

6. Conclusion

In summary, it is understood from the experimental studies that the magnetoelectric coefficient ($\frac{\delta E}{\delta H}$) of a composite can be increased, using individual phases with higher piezoelectric and higher magnetization values. It was found that the soft piezoelectric phase can increase the ME value and higher Zinc doping was found to improve the magnetoelectric response of PZT – CFO based magnetoelectric composite. The coercive field of the CZF particles was lower than the CFO particles, which contributes towards the easy reversal of magnetic domain under AC magnetic field. The addition of Zn, contributes to the increase in saturation magnetization and hence the permeability. It has already reported that with high magnetic permeability inside the ferromagnetic material, magnetic flux strength increases and hence it helps to increase the effective piezomagnetic coefficient ($d\lambda/dH$). This increase in piezomagnetic coefficient helps to improve the magnetoelectric coefficient of the composites.

Acknowledgements

I would like to thank Prof. Shashank Priya of Penn State University for his technical guidance.

Author details

Rashed A. Islam
Alphabet Inc., 1600 Amphitheatre Parkway, Mountain View, CA 94043, USA

*Address all correspondence to: rashed_ms@yahoo.com

IntechOpen

© 2021 The Author(s). Licensee IntechOpen. This chapter is distributed under the terms of the Creative Commons Attribution License (http://creativecommons.org/licenses/by/3.0), which permits unrestricted use, distribution, and reproduction in any medium, provided the original work is properly cited.

References

[1] Hur N, Park S, Sharma PA, Ahn JS. S. Guha and S-W Cheong. Nature. 2004; **429**:392

[2] Wang J, Zheng H, Lofland SE, Ma Z, Ardabili LM, Zhao T, et al. Science. 2004;**303**:661

[3] Ederer C, Spaldin N. Nature Materials. 2004;**3**:849

[4] Eerenstein W, Mathur ND, Scott JF. Nature. 2006;**442**:759

[5] Kimura T, Goto T, Shintani H, Ishizaki K, Arima T, Tokura Y. Nature. 2003;**426**:55

[6] Lottermoser T, Lonkal T, Amann U, Hohlwein D. J. Ihiringer and M. Fiebig. 2004;**430**:541

[7] Astrov DN. Sov. Phys. JETP. 1960;**11**: 708

[8] Wang J, Neaton JB, Zheng H, Nagarajan V, Ogale SB, Liu B, et al. Science. 2003;**299**:1719

[9] Van Aken BB, Palstra TTA, Filippetti A, Spaldin NA. Nat. Mater. 2004;**3**:164

[10] Duan CG, Jaswal SS, Tsymbal EY. Phys. Rev. Lett. 2006;**047201**:97

[11] Fiebig M, Lottermoser T, Frohllch D, Goltsev AV, Pisarev RV. Nature. 2002;**419**:818

[12] Zheng M, Wan JG, Wang Y, Yu H, Liu JM, Jiang XP, et al. J. Appl. Phys. 2004;**95**(12):8069

[13] Ryu J, Carazo AV, Uchino K, Kim H. J. Of Electocer. 2001;7:17

[14] Flores VC, Baques DB, Flores DC, Aquino JAM. J. Appl. Phys. 2006;**99**: 08J503

[15] Boomgaard JVD. A.M.J.G. Van Run and J.V Suchtelen. Ferroelectrics. 1976; **10**:295

[16] Boomgaard JVD, Born RAJ. J. Mater. Sci. 1978;**13**:1538

[17] Boomgaard JVD, Terrell DR, Born RAJ, Giller HFJI. J. Mater. Sci. 1974;**9**:1705

[18] Van Run AMJG, Terrell DR, Scholing JH. J. Mater. Sci. 1974;**9**:1710

[19] Lupeiko TG, Lisnevskaya IV, Chkheidze MD, Zvyagintsev BI. Inorg. Mater. 1995;**31**:1139

[20] T. G.Lupeiko, I. B.Lopatina, SS Lopatin, and IP Getman. Neorg. Mater. 1991;**27**(11):2394

[21] Lupeiko TG, Lopatina IB, Kozyrev IV, Derbaremdiker LA. Neorg. Mater. 1991;**28**(3):632

[22] Bokhan YI, Laletin VM. Inorg. Mater. 1996;**32**(5):634

[23] TG Lupeiko, SS Lopatin, IV Lisnevskaya, and BI Zvyagintsev, Inorg. Mater., 30, 1353, (1994).

[24] Dai YR, Bao P, Zhu JS, Wan JG, Shen HM, Liu JM. J. Appl. Phys. 2004; **96**(10):5687

[25] Islam RA, Priya S. J of Mat. Sci. 2008;**43**(4):1497-1500

[26] Islam RA, Priya S. J of Mat. Sci. 2008;**43**(10):3560-3568

Section 2

Driving Principles and Structural Design

Chapter 2

Parasitic Motion Principle (PMP) Piezoelectric Actuators: Definition and Recent Developments

Lin Zhang and Hu Huang

Abstract

Stepping piezoelectric actuators have achieved significant improvements to satisfy the urgent demands on precision positioning with the capability of long working stroke, high accuracy and micro/nano-scale resolution, coupled with the merits of fast response and high stiffness. Among them, inchworm type, friction-inertia type, and parasitic type are three main types of stepping piezoelectric actuators. This chapter is aimed to introduce the basic definition and typical features of the parasitic motion principle (PMP), followed by summarizing the recent developments and achievements of PMP piezoelectric actuators. The emphasis of this chapter includes three key points, the structural optimization, output characteristic analysis and performance enhancement. Finally, the current existing issues and some potential research topics in the future are discussed. It is expected that this chapter can assist relevant researchers to understand the basic principle and recent development of PMP piezoelectric actuators.

Keywords: parasitic motion principle, piezoelectric actuator, long working stroke, flexure hinge-based compliant mechanism, backward motion

1. Introduction

Nowadays, long working stroke precision positioning systems with micro-to-nano resolution are significantly demanded in many scientific studies and industrial fields [1–3]. Most of the conventional actuators can hardly satisfy the requirements on positioning resolution for precision positioning systems, such as hydro-motors, direct/alternating current motors, pneumatic elements, et al., even with the merits of large output capability, fast response, and long working stroke [4–6].

The piezoelectric actuator is one of the potential alternatives for high-resolution precision positioning systems [7–10]. Up to now, various of piezoelectric-driven positioning systems with flexure hinge-based compliant mechanisms have been developed and widely applied in many scientific and industrial applications, such as atomic force microscopy (AFM) [11–13], fast tool servo (FTS) single-point diamond turning [14–16] and optical adaptive mirror [17–19], et al. Generally, restricted by the inverse piezoelectric effect of current piezoelectric materials, the displacement of a single piezoelectric element is limited within tens of nanometers to several micrometers [20]. The applications of such positioning stages are only employed within limited scopes due to micro-scale working stroke. In order to extend the working stroke of piezoelectric elements, several methods have been proposed and investigated [21–23], which can be classified according to the motion principle into the direct-driven principle, ultrasonic principle, and stepping principle. Direct-driven principle is the initial application in piezoelectric actuators. With the assistance of flexure hinge-based compliant mechanisms, it is found that the working stroke can be amplified up to several times of the original displacement of a single piezoelectric element. The maximum working stroke is extended to tens of micrometers [24–26]. However, it is still not long enough for most of the applications, and furthermore complicated flexure hinge-based compliant mechanisms deteriorate the static and dynamic characteristics of the piezoelectric actuators, reducing structural stiffness and intrinsic resonant frequency. Therefore, the direct-driven principle gradually loses its popularity in the recent years. Ultrasonic principle utilizes the resonance of stators to drive the slider/rotor. However, the interfacial wear and heat generation are lack of adequate solution to date, especially in high-speed & full-load motion [27, 28]. Stepping principle realizes the long working stroke by step displacement accumulation. By this way, high-precision positioning accuracy can be achieved in long working stroke. Hence, stepping principle has attracted much attention in the piezoelectric actuator development in the recent decades.

Various of stepping piezoelectric actuators can be further classified into three motion types, involving inchworm type, friction-inertia type, and parasitic type [3, 29–31]. Inchworm type, as a kind of bionic driving type, mimics the motion principle of inchworms in nature, which alternates the clamping and driving units to move forward and backward. Thus, its control strategy, structural assembly and the motion sequence are generally complicated. Friction-inertia type refers to a kind of spontaneous jerking motion that can occur, while two mass blocks alternate between sticking to each other and sliding over each other, with a corresponding tuning the friction and inertia forces. Compared with the inchworm type, the basic structure and control system of friction-inertia type are largely simplified but associated with loss on loading capability.

Parasitic type is a new solution to acquire both long working stroke and large output capability by adopting the parasitic motion of flexure hinge-based compliant mechanisms, which is commonly restricted in previous designs [32, 33]. Up to now, tens of PMP piezoelectric actuators based on various of flexure hinge-based compliant mechanisms have been developed with great success and achievement on improving working stroke and output capability. The purpose of this chapter is to introduce the basic parasitic motion principle, review the developments and achievements in recent years, and finally point out some potential issues and current challenges in this research.

2. Introduction to the parasitic motion principle

Different from other kinds of motion principles, the parasitic principle belongs to a kind of dependent motion, which generally accompanies with an independent motion, as illustrated in **Figure 1(a)**. When a load F is applied at the end of a cantilever beam, it will be bent with two motion components in x and y directions. The motion component in y direction is the major motion, which is directly induced by the load F, while the motion component in x direction is called as the parasitic motion. It simultaneously occurs with the major motion, which is generally regarded as an undesired motion component in previous studies. In general, the parasitic motion is much smaller than the major motion, but this dependent motion may deteriorate positioning accuracy and lead to more issues in calibration. On the

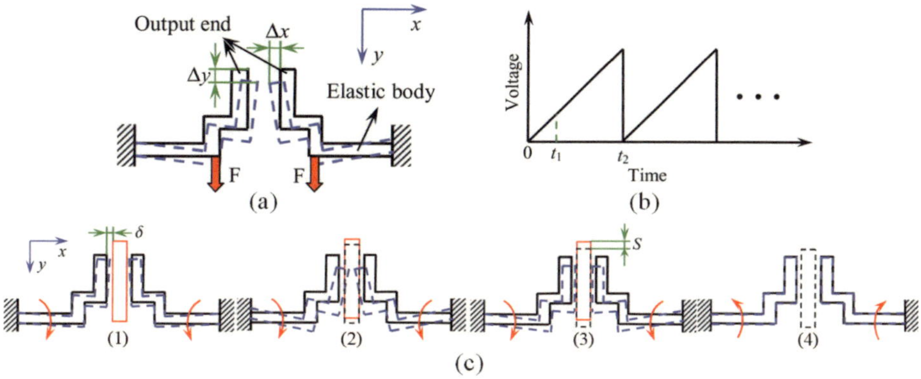

Figure 1.
Schematic diagrams of the parasitic motion principle: (a) generation of parasitic motion when bending a cantilever, (b) saw-tooth wave control signal, and (c) motion principle of the PMP piezoelectric actuator in one step [23].

other hand, if the parasitic motion of flexure hinge-based compliant mechanisms can be appropriately adopted in the design of piezoelectric actuators, it can be employed as a motion task by utilizing lower degree of freedom (DOF) with easier control, lower cost, less complexity of kinematics and simple structure. By employing specially designed control signal, for instance, the saw-tooth wave as shown in **Figure 1(b)**, applied to the piezoelectric element, the relative displacement is realized, and thus the stepping motion is achieved. Therefore, the PMP piezoelectric actuator becomes popular since its emergence in recent years.

Figure 1(c) shows the motion principle in one step of the PMP piezoelectric actuator. This kind of actuators is generally consisted of two sections, the stator and the slider/rotor. At the initial step (1), the stator and the slider are in separated state with an initial gap δ between each other. Then, in step (2), with a moment/force slowly applied to the flexure hinge-based compliant mechanism, the initial gap δ is filled, leading to an initial contact between the stator and the slider/rotor. Afterwards, in step (3), both the major motion and parasitic motion increase with deformation of the flexure hinge-based compliant mechanism. The slider/rotor moves in the same direction with the parasitic motion. Finally, after the slider/rotor moves to the forward displacement/angle in one step, the moment/force is suddenly removed, and the flexure hinge-based compliant mechanism recovers to its initial state and gets ready for the next cycle. In this process, as the stator still contacts with the slider/rotor, a backward motion would generally appear in the final step. Therefore, the PMP piezoelectric actuator could move with one-step displacement ΔS, the one-step forward displacement minus the backward motion. By cycling from step (1) to step (4), the long working stroke can be easily achieved.

3. Similarities and differences with other stepping principles

Inchworm type, friction-inertia type, and parasitic type are three main kinds of motion types in stepping principle to realize long working stroke. Inchworm type, as a kind of bionic principle, employs the driving units and clamping units to obtain long working stroke. The utilization of clamping units facilitates the enhancement on output capability for piezoelectric actuators. In general, the inchworm type actuator consists of three separate parts, one driving unit and two clamping units. The moving processes of the inchworm type piezoelectric actuator are presented in **Figure 2**.

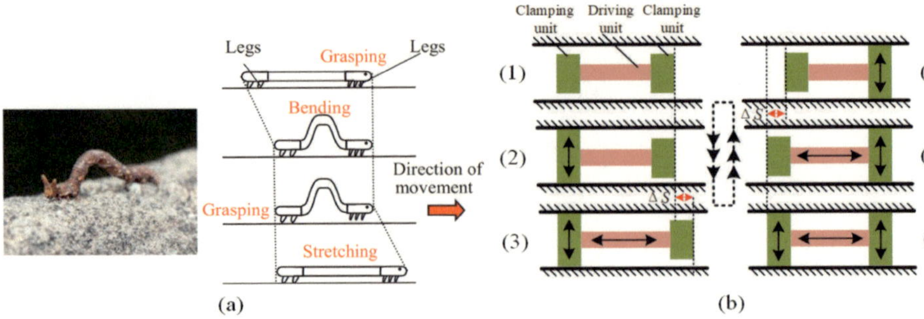

Figure 2.
Motion principle of the inchworm type actuator: (a) moving principle of inchworm in nature [34], and (b) bionic stepping motion principle of inchworm type piezoelectric actuators.

Figure 3 shows the schematic diagram of friction-inertia type motion principle. The motion principle for the friction-inertia type follows the law of momentum conservation. A piezoelectric stack or piezoelectric bimorph, between two objects with different weights, is driven by a special control signal. At the initial step (1), the piezoelectric element is in its original status and connects two blocks. Then, in the step (2), the piezoelectric element extends gradually with the increase of driving voltage, and one block follows the movement of the piezoelectric element due to the static friction. In this process, there is no relative motion between the two objects. Afterwards, in step (3), the driving voltage suddenly drops to zero and the piezoelectric element loses power. It quickly recovers to the initial status, but the moving block remains in its position due to the inertial force. Following these steps, a small displacement occurs in this process. Based on the moving process, the friction-inertia actuator involves two motion types: impact-friction type and stick–slip type [3]. The main difference from impact-friction type is that, in stick–slip type, one end of the driving element is connected to the base and the other end drives the mass block by surface friction.

These three motion principles have some similarities and differences. According to the previous research, the performance comparison of these three motion types of stepping principle piezoelectric actuators is listed in **Table 1**. From the list, the inchworm type piezoelectric actuators dominate the high resolution and large output capability, but the free-load motion velocity is lower than its counterparts. Whereas, the friction-inertia type and parasitic type piezoelectric actuators have superiorities on motion speed and control system but deficiency on the output capability.

Compared with the inchworm type piezoelectric actuator, the structure of the PMP piezoelectric actuator is compact and its control strategy is quite simple. The parasitic motion completes the actions of clamping and driving in inchworm type motion. The re-clamping in the inchworm type is neglected in the parasitic type motion. Therefore, the difficulty and complexity on control system drop down but the output load capability is sacrificed to some extent. As a similar motion like friction-inertia type, the main difference is on the interaction between the driver and the slider/rotor. In PMP piezoelectric actuators, the normal clamping force, as well as the friction, between the driver and the slider/rotor becomes large as the voltage increases, while the forces are generally maintained the same in friction-inertia type piezoelectric actuator. All in all, the parasitic motion principle can be treated as a combination of inchworm principle and friction-inertia principle to some extent. It simplifies the structures and control strategy of inchworm principle, and exceeds the output capability of friction-inertia motion principle by increasing the clamping force.

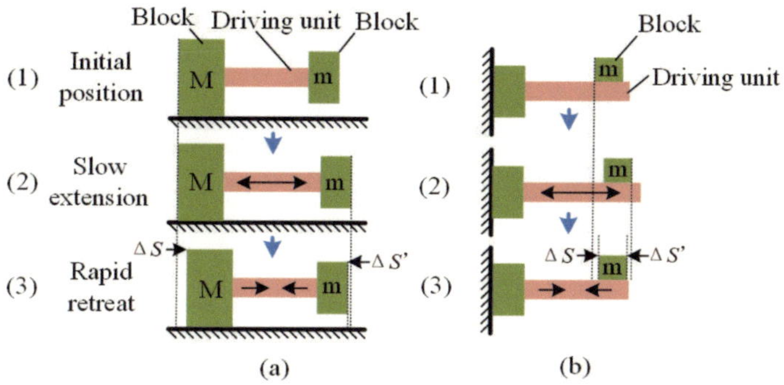

Figure 3.
Motion principles of two friction-inertia types actuators: (a) impact-drive type, and (b) stick–slip type [3].

Type	Positioning resolution	Carrying capability	Motion speed	Control strategy
Inchworm	High	High	Slow	Complex
Friction-inertia	Middle	low	Fast	Simple
Parasitic	Middle	Middle	Fast	Simple

Table 1.
Performance comparison of three motion types of stepping principle piezoelectric actuators.

4. Developments and achievements

In 2012, Huang et al. was the first one proposing the PMP piezoelectric actuator by using two microgrippers [23]. The three-dimension (3D) model of the actuator is shown in **Figure 4(a)**. A slider is parallelly placed between two elastic clampers. In most cases, the micro-gripper is employed to precisely manipulate micro/nano-scale objects. However, with the major motion Δx clamping the objects, a parasitic motion Δy pulls the slider to move a minor distance being vertical to the clamping direction. Driven by the saw-tooth wave, a long working stroke was accumulated by step-by-step motion. Various of experiments were conducted with 25 V ~ 100 V driving voltages and 1 Hz ~ 5 Hz driving frequencies to prove the practicability of the proposed driving mechanism. In another research, as shown in **Figure 4(b)**, a more compact linear parasitic motion positioning stage consisting of one compact micro-gripper and one piezoelectric element was developed by Huang et al. [35]. The experiments indicated the linear positioning stage can achieve forward and reverse movements with different driving saw-tooth waves, as well as movement velocities and stepping displacement.

By utilizing various of flexure hinge-based compliant mechanisms, some novel kinds of piezoelectric actuators based on parasitic motion are developed. **Figure 5** illustrates novel PMP piezoelectric actuators with bridge-type flexure hinge-based compliant mechanism. This type of flexure hinge-based compliant mechanism is a novel kind of structure used in piezoelectric actuators, which not only amplifies the output displacement but generates coupled motion component as well. The motion principle of the bridge-type flexure hinge-based compliant mechanism is shown in **Figure 5(a)**. Li et al. introduced both linear and rotary PMP piezoelectric actuators based on such mechanism [36, 37], as shown in **Figure 5(b)** and **(c)**. The parasitic motion of the bridge-type flexure hinge-based compliant mechanism was theoretically analyzed and numerically simulated by the elastic-beam theory (EBT), rigid-body method (RBM) and finite element method (FEM), respectively. Dual-servo

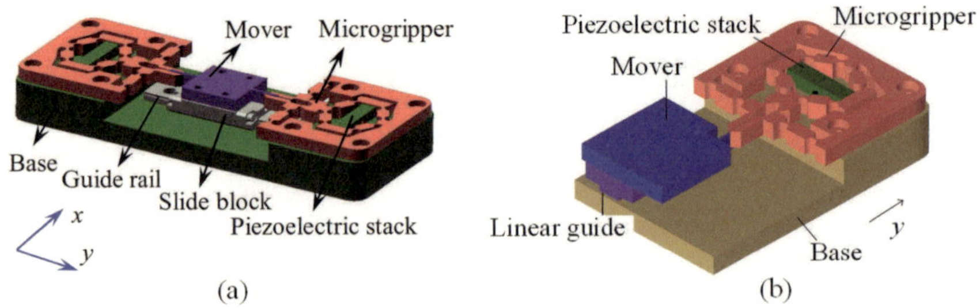

Figure 4.
PMP piezoelectric actuators proposed by Huang et al. [23, 35]. (a) using two microgrippers and (b) using only one microgripper.

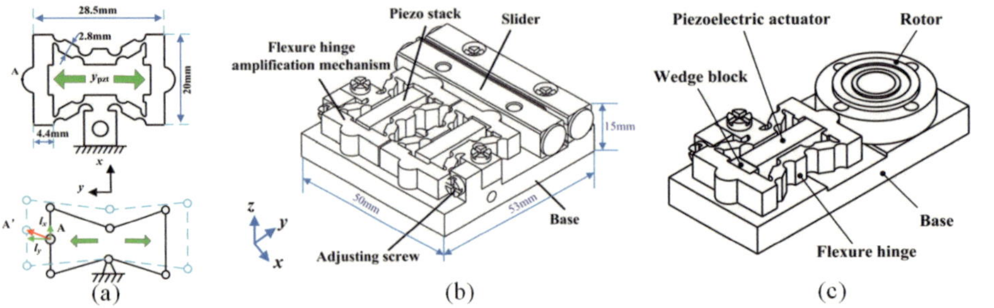

Figure 5.
PMP piezoelectric actuators designed by using the bridge-type flexure hinge-based compliant mechanism: (a) working principle, (b) bidirectional linear actuator by Li et al. [37], and (c) rotary actuator by Li et al. [36].

control strategy was introduced to achieve long working stroke and nano-scale resolution positioning within one single step. Experiments showed that the maximum velocity of 7.95 mm/s was achieved for the linear actuator with the driving voltage of 100 V at a driving frequency of 1000 Hz, while the rotary actuator can reach 32000 μrad/s with the driving voltage of 100 V at a driving frequency of 100 Hz. Wang et al. proposed a bidirectional complementary-type actuator, which utilized parasitic motion in the longitudinal deformation for driving and clamping [38]. Compared with the current existing prototypes, it reduced the motion coupling to 4%, and optimized the step consistency and driving capability to a large extent.

After that, several different PMP piezoelectric actuators are proposed by employing different flexure hinge-based compliant mechanisms, i.e. asymmetric flexure hinge-based compliant mechanism, parallelogram flexure hinge-based compliant mechanism and trapezoid flexure hinge-based compliant mechanism. In comparison with the bridge-type flexure hinge-based compliant mechanism, the asymmetric flexure hinge-based compliant mechanism has simple structure with high stiffness. Li et al. proposed an asymmetric flexure hinge-based compliant mechanism, as shown in **Figure 6(a)**, to amplify the parasitic motion in the PMP piezoelectric actuator [39]. By introducing the asymmetric flexure hinge-based compliant mechanism, the resolution of the proposed linear PMP piezoelectric actuator was improved to 0.68 μm. The maximum speed can reach 4.676 mm/s and the maximum output load was enhanced to 91.3 g. Another linear actuator was proposed by Li et al., as shown in **Figure 6(b)**. The lever-type piezoelectric actuator could achieve bidirectional motion driven by a single piezoelectric element [40]. Under the symmetry of 20% and 80%, the maximum forward velocity was 7.69 mm/s and maximum reverse velocity was 7.12 mm/s, respectively. Gao et al. presented a PMP piezoelectric actuator based on an asymmetrical flexure hinge-based compliant mechanism [41], as shown in **Figure 6(c)**. The authors designed

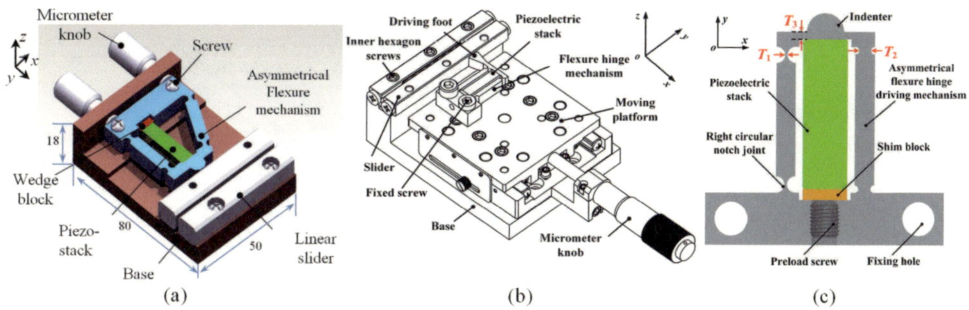

Figure 6.
PMP piezoelectric actuators with the asymmetric flexure hinge-based compliant mechanism developed by (a) Li et al. [39], (b) Li et al. [40], and (c) Gao et al. [41].

four bars with different thickness right-circle flexure hinges to achieve improvement on output speed and efficiency. Simulations were employed to optimize the structure parameters and the experimental results indicated that the maximum velocity of the proposed piezoelectric actuator reached 15.04 mm/s under the driving voltage of 100 V at a driving frequency of 490 Hz.

Parallelogram flexure hinge-based compliant mechanism is another widely used structure in PMP piezoelectric actuators. Due to its simple structure and flexible design, it gains popularity in studies. Li et al. first introduced the parallelogram flexure hinge-based compliant mechanism in the PMP piezoelectric actuators and characterized the performance of the proposed actuator [42], as shown in **Figure 7(a)**. In the case, the maximum free-load motion speed of the proposed PMP piezoelectric actuator was 14.25 mm/s under the driving voltage of 100 V at a driving frequency of 2000 Hz. Some modified parallelogram structures were also proposed with improved driving capability by Li et al. [43, 44]. By combining the parallelogram flexure hinge-based compliant mechanism and asymmetrical flexure hinge-based compliant mechanism, several different PMP piezoelectric actuators were developed, as shown in **Figure 7(b)** and **(c)**. The parasitic motion was characterized by EBT and FEM, and the experiments proved the feasibility of the proposed piezoelectric actuator and simplification of walking type for piezoelectric actuators. Furthermore, Gao et al. developed another modified parallelogram flexure hinge-based compliant mechanism in PMP piezoelectric actuators [45]. By adopting different stiffness flexure hinges, parasitic motion displacement was amplified, and the working performance was investigated by a prototype, as shown in **Figure 7(d)**.

The special mechanical properties of the trapezoid flexure hinge-based compliant mechanism attract the attention from researchers. By adjusting the structural parameters, various kinds of trapezoid flexure hinge-based compliant mechanism with different mechanics characteristics can be obtained. Some of them can easily bring in the parasitic motion in the deformation. Li et al. investigated the possibility of introducing trapezoid flexure hinge-based compliant mechanism into PMP piezoelectric actuators [46], and manufactured a prototype to study the kinematic properties of the proposed PMP piezoelectric actuator. The design of the PMP piezoelectric actuator is shown in **Figure 8(a)**. The right-circular flexure hinges with different thickness were employed in the prototype design of the trapezoid flexure hinge-based compliant mechanism, which had the capability to achieve the parasitic motion. The moving process was characterized and verified by theoretical calculation, numerical simulation and experiments. The experimental results indicated that the maximum speed was 180 μm/s with the driving voltage of 100 V at a driving frequency of 220 Hz. Cheng et al. analyzed the trapezoid flexure hinge-based compliant mechanism and applied such structure into the development

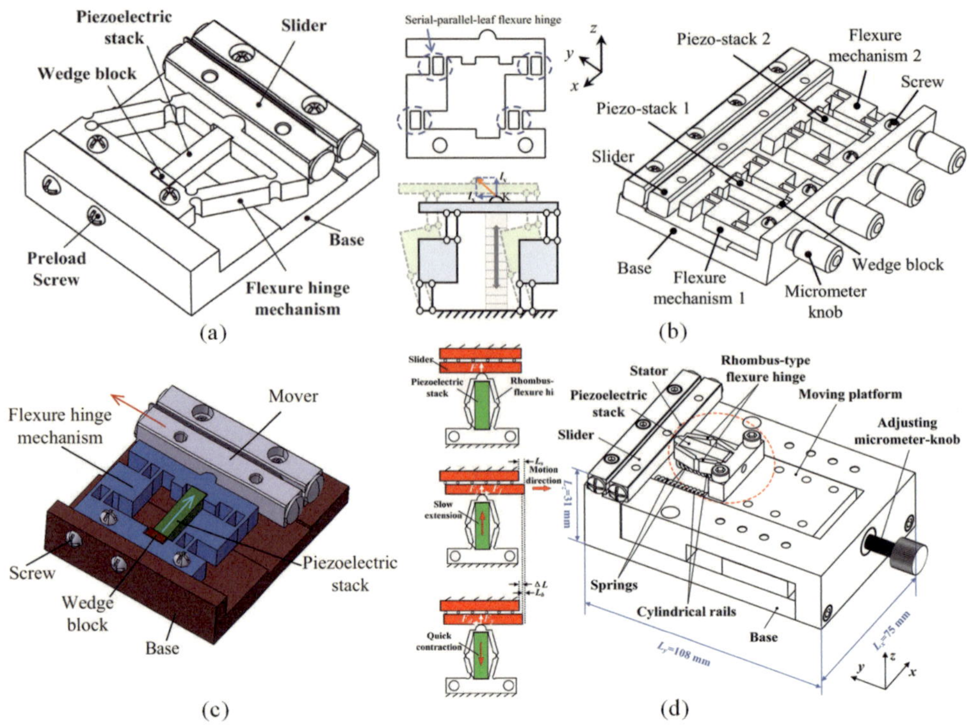

Figure 7.
PMP piezoelectric actuators designed with the parallelogram flexure hinge-based compliant mechanism developed by (a) Li et al. [42], (b) Wen et al. [43], (c) Wan et al. [44], and (d) Gao et al. [45].

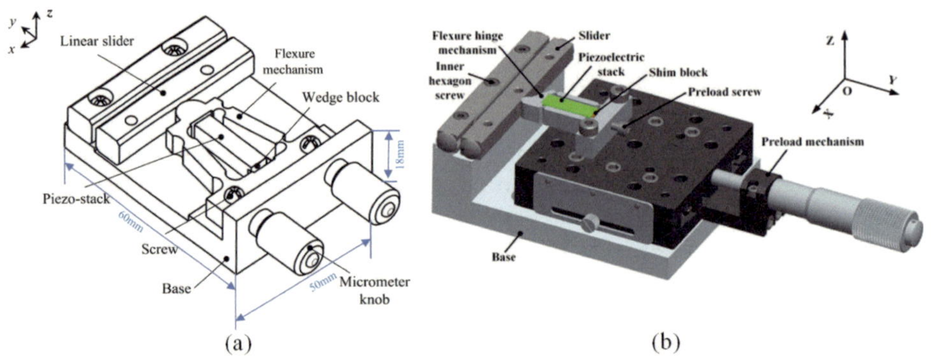

Figure 8.
PMP piezoelectric actuators with trapezoid flexure hinge-based compliant mechanism: (a) equilateral triangle flexure structure by Li et al. [46], and (b) right-circular flexure structure by Cheng et al. [47].

of PMP piezoelectric actuators [47]. They attempted to optimize the asymmetrical flexure hinge-based compliant mechanism to achieve large static friction force in slow extension phase while low kinetic friction force in quick backward phase. The prototype was fabricated to confirm the proposed structure. The maximum speed and maximum output load were 5.96 mm/s and 3 N under the driving voltage of 100 V at a driving frequency of 500 Hz. Another research employing a modified trapezoid flexure hinge-based compliant mechanism was developed by Lu et al. [48], which achieved high speed at lower driving frequency.

Apart from the most used flexure hinge-based compliant mechanism in PMP piezoelectric actuators, some other structures are also introduced to enhance the parasitic motion. The symmetrical flexure hinge-based compliant mechanism was applied into the PMP piezoelectric actuator by Yao et al. [49]. The design of the actuator is shown in **Figure 9(a)**. The structural characteristics and motion displacement were theoretically analyzed and predicted by FEM. The motion principle of the coupled symmetrical flexure hinge-based compliant mechanism is shown in **Figure 9(b)**. With the assistance of the coupled symmetrical flexure hinge-based compliant mechanism, the developed PMP piezoelectric actuator achieved notable improvement on kinematic performance and large output capability. The experiments showed that the minimum step displacement was 0.495 μm under the input driving voltage of 30 V at a driving frequency of 1 Hz and the maximum speed was 992.4 μm/s with the input driving voltage of 120 V at a driving frequency of 400 Hz. Lu et al. developed another kind of coupled symmetrical flexure hinge-based compliant mechanism for linear PMP piezoelectric actuators [50]. The FEM simulation under static load is shown in **Figure 9(c)**. The feasibility of the designed structure was confirmed by the numerical simulation and experiment.

Besides the aforementioned PMP piezoelectric actuators, Li et al. investigated a "Z-shaped" symmetric flexure hinge-based compliant mechanism in the PMP piezoelectric actuator [51]. Since the symmetric flexure hinge-based compliant mechanisms were rotated with an angle of θ = ±20° to the slider, coupled motion could be achieved in *x* and *y* directions. **Figure 10(a)** shows the 3D model of the PMP piezoelectric actuator. In this case, the system statics and kinetic models were established for better understanding the static and dynamic performances of the proposed linear PMP piezoelectric actuator. Furthermore, a triangular structure with flexure hinge-based compliant mechanism was proposed by Zhang et al. [52], as shown in **Figure 10(b)**. Compared to the existing actuators with similar motion principle, the proposed triangular flexure hinge-based compliant mechanism had the capability to amplify the clamping force as well as the driving force. The proposed actuator achieved several times larger driving force and higher free-load motion speed with similar or even lower driven voltage. Besides these linear PMP piezoelectric actuators, several kinds of rotary PMP piezoelectric actuators with triangular structure were proposed by Zhang et al. to confirm the possibility of the proposed flexure hinge-based compliant mechanism in PMP piezoelectric actuators [53]. To enhance the load capability for both forward and backward motions, a shared driving foot flexure hinge-based compliant mechanism, equipped with two piezoelectric stacks, was proposed by Zhang et al. [54]. The 3D model and the working principle are shown in **Figure 10(c)**. Experimental results indicated that the actuator could achieve a free-load maximum forward and backward speed up to 18.6 mm/s and 16.0 mm/s, respectively. The output capacity was largely improved to 2.0 kg for the both driving directions. Zhang et al. developed a linear piezoelectric actuator

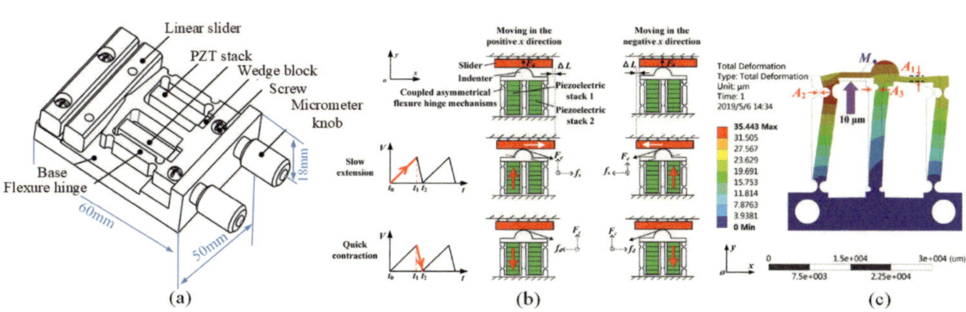

Figure 9.
PMP piezoelectric actuators with coupled symmetrical flexure hinge-based compliant mechanism: (a) linear piezoelectric actuator by Yao et al. [49], (b) motion principle of the symmetrical flexure hinge-based compliant mechanism, and (c) FEM simulation by Lu et al. [50].

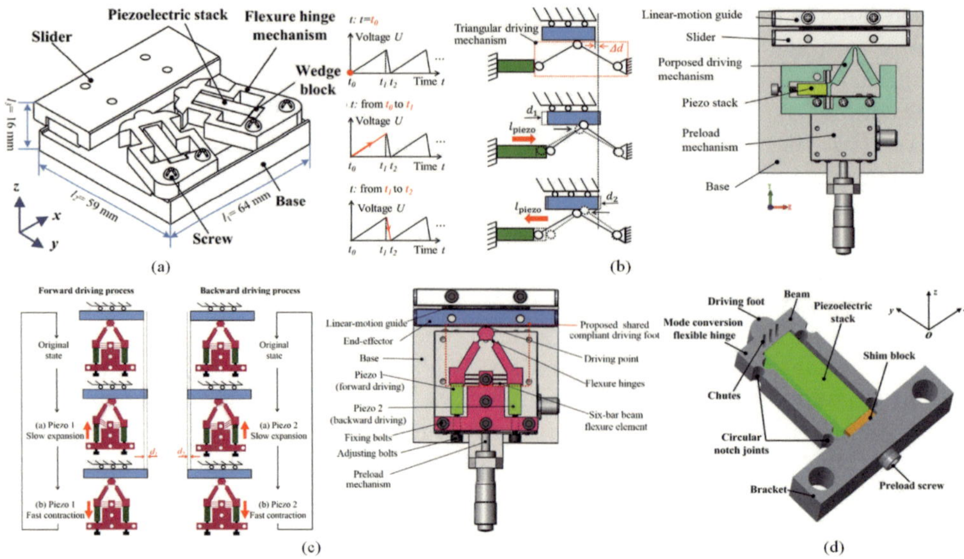

Figure 10.
PMP piezoelectric actuators designed by using (a) a "Z-shaped" flexure hinge-based compliant mechanism by Li et al. [51], (b) triangular-type flexure hinge-based compliant mechanism by Zhang et al. [52], (c) shared driving foot mechanism by Zhang et al. [54], and (d) mode conversion flexure hinge-based compliant mechanism by Zhang et al. [55].

with mode conversion flexure hinge-based compliant mechanism [55], as shown in **Figure 10(d)**. The mode conversion flexible hinge with a structure of chutes achieved lateral motion and constant phase difference with symmetrical waveform. Different parameters of the chutes were analyzed by FE simulation and experiment. The experimental results showed good agreement with the simulation analysis.

More recently, some compact flexure hinge-based compliant mechanisms are introduced into the PMP piezoelectric actuators to enhance the performances. Wang et al. reported a rotary piezoelectric actuator with centrosymmetric flexure hinge-based compliant mechanism [56]. The structure of the proposed piezoelectric actuator is presented in **Figure 11(a)**. The motion principle was analyzed by FEM, which was further confirmed by the experiment. Both the output capability and moving resolution of the proposed actuator were improved, and the clockwise and anticlockwise rotations can be switched by adjusting the driving voltage waveform. Besides the rotary PMP piezoelectric actuator, another linear PMP piezoelectric actuator was then introduced to confirm the feasibility of bidirectional PMP piezoelectric actuator [57]. The structure of the bidirectional piezoelectric actuator is illustrated in
Figure 11(b). Furthermore, by employing two lever-type flexure hinge-based compliant mechanism, Li et al. developed a 2-DOF piezoelectric-driven precision positioning stage by using parasitic motion [58]. As shown **Figure 11(c)**, the stage consisted of two layers with the same driven structures and the L-shape flexure hinges made the structure compact with piezoelectric stacks being parallel to the slider. The prototype achieved relatively large output displacement over 1,600 μm with good linearity. Wang et al. developed a high-velocity rotary parasitic type piezoelectric positioner [59]. A compact rotational symmetric flexure mechanism with self-centering function was employed to generate parasitic motion to drive the rotor, as shown in **Figure 11(d)**. The experimental results showed the proposed positioning stage achieved the maximum speed of 151.4 mrad/s, which was much greater than most of the current reported non-resonant piezoelectric positioner.

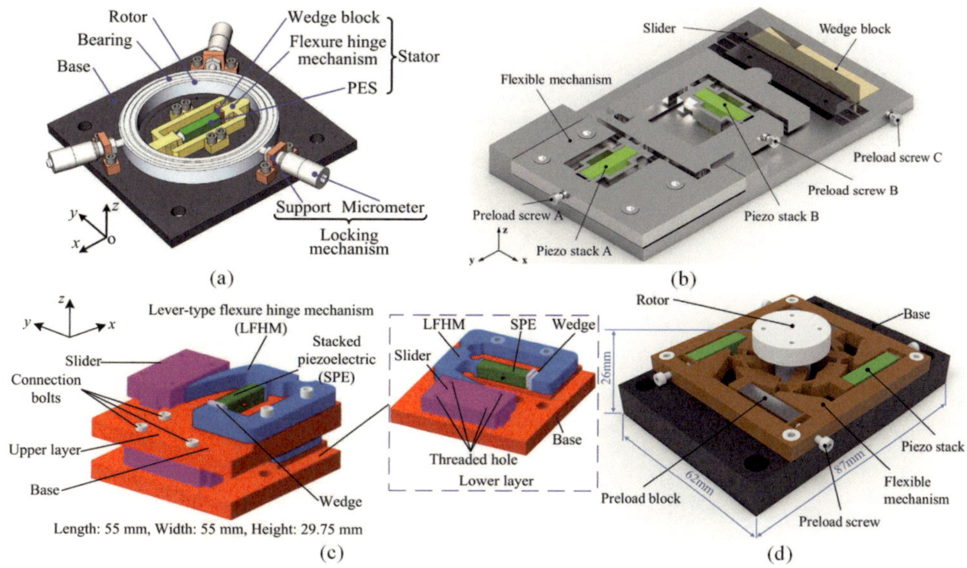

Figure 11.
PMP piezoelectric actuators designed by using (a) centrosymmetric flexure hinge-based compliant mechanism for rotary actuator [56], (b) two lever-type flexure hinge-based compliant mechanism for linear actuator [57], (c) "L-shape" compact 2-DOF actuator [58], and (d) rotational symmetric flexure hinge-based compliant mechanism for rotary actuator [59].

In order to obtain better understanding of the motion characteristics, some in-depth research is conducted to clarify the nature in some phenomena, such as backward motion and interfacial interaction. Huang et al. firstly investigated the non-linearity and backward motion in one step of a rotary PMP piezoelectric actuator [60], as shown in **Figure 12(a)**. The analysis indicated that the non-linearity in one step was due to the fit-up gap of the bearing and the self-deformation of the flexible micro-gripper when contacted with the slider, while the backward motions was attributed to the non-ideal driving wave. Furthermore, the characteristics of a linear PMP piezoelectric actuator were also investigated [61], and a dynamics model was provided for system control and optimization. Taking some potential factors, such as the coupling angle, the driving signal symmetry, the mover mass and the preload force, into consideration, the model analyzed the influences of these factors on the output, such as the step length, the backward ratio and the maximum load. Based on the characterization and analysis of the PMP piezoelectric actuators, some strategies were introduced to suppress the backward motion. Huang et al. employed two piezoelectric stacks to realize the synergic motion principle [62]. One of the piezoelectric stacks was used for driving and the other was used for lifting, as shown in **Figure 12(b)**. By theoretical analysis and experiments, the actuator could achieve stepping displacement without backward motion with the aid of synergic driving principle. Another strategy on suppression of backward motion in PMP piezoelectric actuators was by means of the sequential control method [63]. As shown in **Figure 12(c)**, two flexure-based hinge mechanisms with different displacement amplification rates in x and y directions were responsible for driving and lifting, respectively. Compared with some conventional PMP piezoelectric actuators, the backward motion was suppressed under the sequential control method.

Up to now, more detailed phenomena in PMP piezoelectric actuators are focused and analyzed to enhance the performances. Wang et al. investigated the influence of initial gap on the one-stepping characteristics of PMP piezoelectric actuators

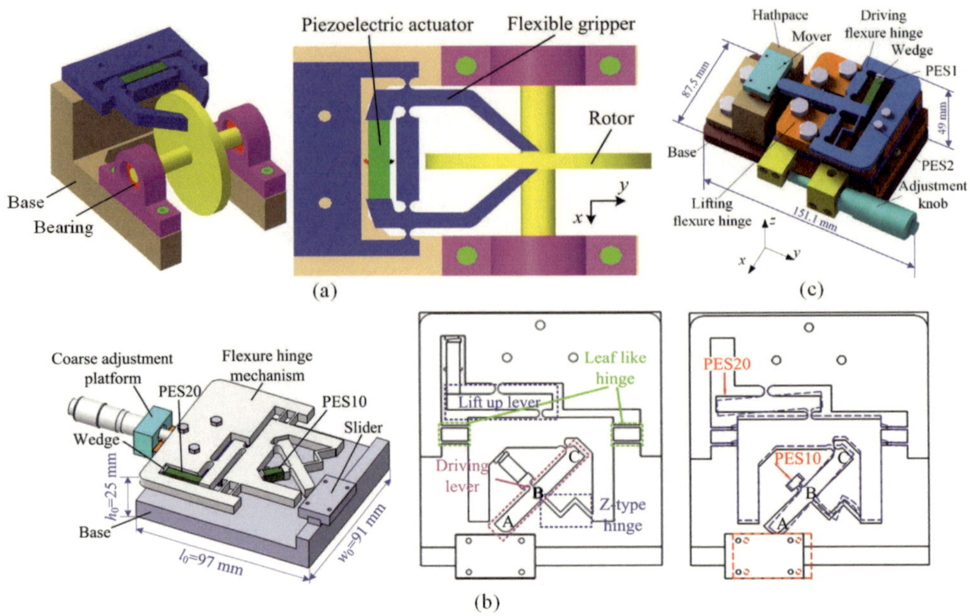

Figure 12.
Mechanism investigations and further improvement on (a) non-linear and backward motion in rotary actuator [60], (b) synergic motion principle by two piezoelectric stacks [62], and (c) sequential control method to suppress the backward motion [63].

[64]. The experimental results showed that the initial gap significantly affected the output characteristics. As shown in **Figure 13(a)**, the previous sudden return (backward motion) transformed into sudden jump, and between them, there was a transition stage, i.e. smooth motion. Another study on preloading was conducted by Yang et al. [65]. By varying the preloading between the flexure hinge-based compliant mechanism and slider, the piezoelectric actuator worked under two different motion modes. Under the new motion mode, the output performances were studied with different initial gaps, driving voltages, driving frequencies, and vertical loads. In addition, the contact force was also measured in PMP piezoelectric actuator by Xu et al. [66], as shown in **Figure 13(b)**. Since the contact force has never been quantitatively detected, it is difficult for keeping the performance uniformity of such actuator in previous studies. By integrating a cantilever beam into the driving unit for measuring the contact force, the actuator could optimize the loading capacity and motion stability by adjusting driving voltage and frequency. The experiments verified the feasibility, and the corresponding actuator was applicable.

Parasitic type piezoelectric actuator is a novel member in the family of stepping actuators. Thus, there is still a lot of research to be done to make the underlying mechanism clear, optimize the structure & control strategy, and enhance the output performances. Although several potential issues have been solved and some achievements have been obtained, the PMP piezoelectric actuators are still far from mass production and wide applications in industry. For example, the nature of the interfacial interaction, compact & simple structures to suppress the backward motion and many related issues are still the stumbling blocks on the way to completion.

5. Issues and future directions

With the introduction of stepping motion principle into piezoelectric actuators, positioning systems are capable to achieve long working stroke and micro-to-nano

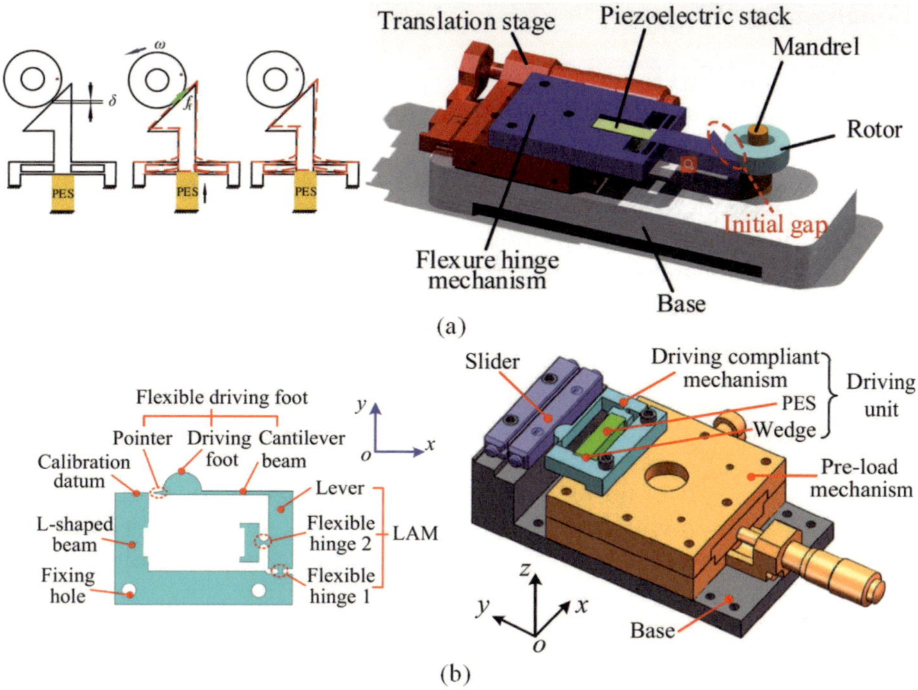

Figure 13.
Mechanical and mechanism investigations on (a) initial gap for one-stepping characteristics [64], and (b) measuring the contact force [66].

positioning resolution. Three motion types of stepping piezoelectric actuators are mostly utilized, inchworm type, friction-inertia type and parasitic type. As one of the most important types, parasitic type showcases the flexibility and massive potential in practical applications in future research and industry. In comparison with the inchworm type piezoelectric actuator, the structure and control strategy of the system are simpler, and it is much easier to obtain high free-load speed. Therefore, further research and efforts should be made to overcome the existing issues in PMP piezoelectric actuators, i.e. backward motion, to satisfy the requirements from general and specific applications, and enhance their adaptation in different conditions.

For the PMP piezoelectric actuators, which have superiorities on simple structure and control system, the low output load and intrinsic backward motions are long-existing issues due to the motion principle. Although some studies attempt to address these issues, some other issues come with the solution. For example, the suppression of backward motion came with increasing complexity of structure and control system. It is now still far from the complete to overcome these issues. Therefore, the studies on improvement of output capability and deep understanding on suppression of backward motion should be further conducted. Furthermore, since the relative motion exists in the parasitic type motion, the wear and tear damages can not be neglected, which will reduce the reliability and stability of the actuator in service. So, the deep understanding and optimization of the interfacial interaction between the flexure and the slider/rotor is another topic in future research. Finally, the multi-direction, integration and minimization of PMP piezoelectric actuators become vital for future applications. Only those which combine long stroke, large load, compact size and integrated system will gain popularity in the future precision-actuator market.

6. Conclusions

This chapter reviews the recent developments and achievements on PMP piezoelectric actuators. Combined with stepping motion principle, the PMP piezoelectric actuator acquires the capabilities on long working stroke and relatively large output capability, which breaks through the long-standing obstacles on micrometric working stroke of single piezoelectric element. In addition, some novel flexure-based hinge mechanisms are introduced to enhance the performances of the parasitic type motion, which not only extend the motion displacement in one step but also improve the motion stability in long working stroke. In addition, the underlying potential issues, i.e. backward motion and contact force, are investigated to understand the nature of the mechanism. By utilizing theoretical analysis and FE simulation, novel structures and driving strategies are applied to suppress the backward motion and improve the motion speed by adjusting interfacial interaction. These prototypes demonstrate better performances than previous parasitic type actuators, verifying the feasibility of the proposed methods. However, further studies should be conducted for improving the performances and overcoming current issues to satisfy the increasing demands for precision positioning and related applications.

Acknowledgements

This work was supported by the National Natural Science Foundation of China (Grant No. 52075221), the Young Elite Scientists Sponsorship Program by CAST (YESS) (Grant No. 2017QNRC001), and the Fundamental Research Funds for the Central Universities (2019-2021).

Conflict of interest

The authors declare no conflict of interest.

Author details

Lin Zhang[1] and Hu Huang[2*]

1 Department of Mechanical Engineering, Keio University, Yokohama, Japan

2 Key Laboratory of CNC Equipment Reliability, Ministry of Education, School of Mechanical and Aerospace Engineering, Jilin University, Changchun, Jilin, China

*Address all correspondence to: huanghu@jlu.edu.cn

IntechOpen

© 2021 The Author(s). Licensee IntechOpen. This chapter is distributed under the terms of the Creative Commons Attribution License (http://creativecommons.org/licenses/by/3.0), which permits unrestricted use, distribution, and reproduction in any medium, provided the original work is properly cited.

References

[1] Zhang X and Xu Q. Design, fabrication and testing of a novel symmetrical 3-DOF large-stroke parallel micro/nano-positioning stage. Robotics and Computer-Integrated Manufacturing 2018; 54: 162-172. DOI: https://doi.org/10.1016/j.rcim.2017.11.006.

[2] Fleming AJ. A review of nanometer resolution position sensors: Operation and performance. Sensors and Actuators A: Physical 2013; 190: 106-126. DOI: https://doi.org/10.1016/j.sna.2012.10.016.

[3] Li J, Huang H and Morita T. Stepping piezoelectric actuators with large working stroke for nano-positioning systems: A review. Sensors and Actuators A: Physical 2019; 292: 39-51. DOI: https://doi.org/10.1016/j.sna.2019.04.006.

[4] D. S, A. P and D. P, et al. A New Type of Motor: Pneumatic Step Motor. IEEE/ASME Transactions on Mechatronics 2007; 12: 98-106. DOI: 10.1109/TMECH.2006.886258.

[5] J. K, S. C and K. C, et al. Position Estimation Using Linear Hall Sensors for Permanent Magnet Linear Motor Systems. Ieee Transactions On Industrial Electronics 2016; 63: 7644-7652. DOI: 10.1109/TIE.2016.2591899.

[6] Gradl C, Plöckinger A and Scheidl R. Sensorless position control with a hydraulic stepper drive — Concept, compression modeling and experimental investigation. Mechatronics 2016; 35: 91-101. DOI: https://doi.org/10.1016/j.mechatronics.2016.01.004.

[7] Li J, Zhao H and Qu H, et al. A piezoelectric-driven rotary actuator by means of inchworm motion. Sensors and Actuators A: Physical 2013; 194: 269-276. DOI: https://doi.org/10.1016/j.sna.2013.02.020.

[8] Li J, Zhao H and Shao M, et al. Design and experimental research of an improved stick–slip type piezo-driven linear actuator. Advances in Mechanical Engineering 2015; 7: 757550584. DOI: 10.1177/1687814015595016.

[9] Li J, Zhao H and Shao M, et al. Design and experiment performances of an inchworm type rotary actuator. Review of Scientific Instruments 2014; 85: 85004. DOI: 10.1063/1.4892998.

[10] Zhao H, Fu L and Ren L, et al. Design and experimental research of a novel inchworm type piezo-driven rotary actuator with the changeable clamping radius. Review of Scientific Instruments 2013; 84: 15006. DOI: 10.1063/1.4788736.

[11] M. WF, S. ORM and A. JF. Q Control of an Atomic Force Microscope Microcantilever: A Sensorless Approach. Journal of Microelectromechanical Systems 2011; 20: 1372-1381. DOI: 10.1109/JMEMS.2011.2168809.

[12] Zhao J, Gong W and Cai W, et al. Piezoelectric bimorph-based scanner in the tip-scan mode for high speed atomic force microscope. Review of Scientific Instruments 2013; 84: 83706. DOI: 10.1063/1.4818976.

[13] Ruppert MG, Moore SI and Zawierta M, et al. Multimodal atomic force microscopy with optimized higher eigenmode sensitivity using on-chip piezoelectric actuation and sensing. Nanotechnology 2019; 30: 85503. DOI: 10.1088/1361-6528/aae40b.

[14] Syahputra HP, Ko TJ and Chung BM. Development of 2-axis hybrid positioning system for precision contouring on micro-milling operation. Journal of Mechanical Science and

Technology 2014; 28: 691-697. DOI: 10.1007/s12206-013-1132-5.

[15] Zhu Z, Zhou X and Liu Z, et al. Development of a piezoelectrically actuated two-degree-of-freedom fast tool servo with decoupled motions for micro–/nanomachining. Precision Engineering 2014; 38: 809-820. DOI: https://doi.org/10.1016/j.precisioneng.2014.04.009.

[16] Huang P, Wu X and To S, et al. Deterioration of form accuracy induced by servo dynamics errors and real-time compensation for slow tool servo diamond turning of complex-shaped optics. International Journal of Machine Tools and Manufacture 2020; 154: 103556. DOI: https://doi.org/10.1016/j.ijmachtools.2020.103556.

[17] Koyama D, Kashihara Y and Hatanaka M, et al. Movable optical lens array using ultrasonic vibration. Sensors and Actuators A: Physical 2016; 237: 35-40. DOI: https://doi.org/10.1016/j.sna.2015.11.009.

[18] Schneider F, Draheim J and Müller C, et al. Optimization of an adaptive PDMS-membrane lens with an integrated actuator. Sensors and Actuators A: Physical 2009; 154: 316-321. DOI: https://doi.org/10.1016/j.sna.2008.07.006.

[19] Lemke F, Weber PM and Philipp K, et al. Piezo-actuated adaptive prisms for continuously adjustable bi-axial scanning. Smart Materials and Structures 2020; 29: 95004. DOI: 10.1088/1361-665x/ab8a00.

[20] Huang H, Zhao H and Fan Z, et al. Analysis and experiments of a novel and compact 3-DOF precision positioning platform. Journal of Mechanical Science and Technology 2013; 27: 3347-3356. DOI: 10.1007/s12206-013-0856-6.

[21] Chiang M, Development of X-Y Servo Pneumatic-Piezoelectric Hybrid Actuators for Position Control with High Response, Large Stroke and Nanometer Accuracy, Sensors, 2010.10.3390/s100402675

[22] F. W, C. L and Y. T, et al. Design of a Piezoelectric-Actuated Microgripper With a Three-Stage Flexure-Based Amplification. IEEE/ASME Transactions on Mechatronics 2015; 20: 2205-2213. DOI: 10.1109/TMECH.2014.2368789.

[23] Huang H, Zhao H and Yang Z, et al. A novel driving principle by means of the parasitic motion of the microgripper and its preliminary application in the design of the linear actuator. Review of Scientific Instruments 2012; 83: 55002. DOI: 10.1063/1.4711869.

[24] Yao Q, Dong J and Ferreira PM. Design, analysis, fabrication and testing of a parallel-kinematic micropositioning XY stage. International Journal of Machine Tools and Manufacture 2007; 47: 946-961. DOI: https://doi.org/10.1016/j.ijmachtools.2006.07.007.

[25] Y. L and Q. X. A Totally Decoupled Piezo-Driven XYZ Flexure Parallel Micropositioning Stage for Micro/Nanomanipulation. IEEE Transactions On Automation Science and Engineering 2011; 8: 265-279. DOI: 10.1109/TASE.2010.2077675.

[26] Bhagat U, Shirinzadeh B and Clark L, et al. Design and analysis of a novel flexure-based 3-DOF mechanism. Mechanism and Machine Theory 2014; 74: 173-187. DOI: https://doi.org/10.1016/j.mechmachtheory.2013.12.006.

[27] Arabaci T, Cicek Y and Dilsiz A, et al. Influence of tip wear of piezoelectric ultrasonic scalers on root surface roughness at different working parameters. A profilometric and atomic force microscopy study. International Journal of Dental Hygiene 2013; 11:

69-74. DOI: https://doi.org/10.1111/idh.12003.

[28] Shiyang L and Ming Y. Analysis of the temperature field distribution for piezoelectric plate-type ultrasonic motor. Sensors and Actuators A: Physical 2010; 164: 107-115. DOI: https://doi.org/10.1016/j.sna.2010.09.009.

[29] Gao X, Yang J and Wu J, et al. Piezoelectric Actuators and Motors: Materials, Designs, and Applications. Advanced Materials Technologies 2020; 5: 1900716. DOI: https://doi.org/10.1002/admt.201900716.

[30] Zhang ZM, An Q and Li JW, et al. Piezoelectric friction–inertia actuator—a critical review and future perspective. The International Journal of Advanced Manufacturing Technology 2012; 62: 669-685. DOI: 10.1007/s00170-011-3827-z.

[31] Peng Y, Peng Y and Gu X, et al. A review of long range piezoelectric motors using frequency leveraged method. Sensors and Actuators A: Physical 2015; 235: 240-255. DOI: https://doi.org/10.1016/j.sna.2015.10.015.

[32] Mathieu F, Larramendy F and Dezest D, et al. Reducing parasitic effects of actuation and sensing schemes for piezoelectric microelectromechanical resonators. Microelectronic Engineering 2013; 111: 68-76. DOI: https://doi.org/10.1016/j.mee.2013.01.057.

[33] Hwang D, Lee MG and Kim H, et al., Precision XY Positioning Stage With Coupled Flexures and Piezo Actuators for Avoiding Parasitic Tilt Error, 2009, pp. 541-545.10.1115/MSEC2009-84094

[34] Ueno S, Takemura K and Yokota S, et al. Micro inchworm robot using electro-conjugate fluid. Sensors and Actuators A: Physical 2014; 216: 36-42. DOI: https://doi.org/10.1016/j.sna.2014.04.032.

[35] Huang H and Zhao H. Forward and Reverse Movements of a Linear Positioning Stage Based on the Parasitic Motion Principle. Advances in Mechanical Engineering 2014; 6: 452560. DOI: 10.1155/2014/452560.

[36] Li J, Zhou X and Zhao H, et al. Design and experimental tests of a dual-servo piezoelectric nanopositioning stage for rotary motion. Review of Scientific Instruments 2015; 86: 45002. DOI: 10.1063/1.4918295.

[37] J. L, X. Z and H. Z, et al. Development of a Novel Parasitic-Type Piezoelectric Actuator. IEEE/ASME Transactions on Mechatronics 2017; 22: 541-550. DOI: 10.1109/TMECH.2016.2604242.

[38] Wang Y and Yan P. A novel bidirectional complementary-type inchworm actuator with parasitic motion based clamping. Mechanical Systems and Signal Processing 2019; 134: 106360. DOI: https://doi.org/10.1016/j.ymssp.2019.106360.

[39] Li J, Cai J and Wen J, et al. A parasitic type piezoelectric actuator with the asymmetrical trapezoid flexure mechanism. Sensors and Actuators A: Physical 2020; 309: 111907. DOI: https://doi.org/10.1016/j.sna.2020.111907.

[40] Li Y, Li H and Cheng T, et al. Note: Lever-type bidirectional stick-slip piezoelectric actuator with flexible hinge. Review of Scientific Instruments 2018; 89: 86101. DOI: 10.1063/1.5038640.

[41] Gao Q, He M and Lu X, et al. Simple and high-performance stick-slip piezoelectric actuator based on an asymmetrical flexure hinge driving mechanism. Journal of Intelligent Material Systems and

Structures 2019; 30: 2125-2134. DOI: 10.1177/1045389X19862376.

[42] Li J, Zhou X and Zhao H, et al. Design and experimental performances of a piezoelectric linear actuator by means of lateral motion. Smart Materials and Structures 2015; 24: 65007. DOI: 10.1088/0964-1726/24/6/065007.

[43] J. W, N. W and R. W, et al. A Novel Linear Walking Type Piezoelectric Actuator Based on the Parasitic Motion of Flexure Mechanisms. IEEE Access 2019; 7: 25908-25914. DOI: 10.1109/ACCESS.2019.2900381.

[44] Wan N, Wen J and Hu Y, et al. A parasitic type piezoelectric actuator with an asymmetrical flexure hinge mechanism. Microsystem Technologies 2020; 26: 917-924. DOI: 10.1007/s00542-019-04627-5.

[45] Gao Q, Li Y and Lu X, et al. A piezoelectric stick-slip linear actuator with a rhombus-type flexure hinge mechanism by means of parasitic motion. Review of Scientific Instruments 2019; 90: 96102. DOI: 10.1063/1.5082856.

[46] Li J, Chen S and Zhao G, et al. A linear piezoelectric actuator with the parasitic motion of equilateral triangle flexure mechanism. Smart Materials and Structures 2019; 29: 15015. DOI: 10.1088/1361-665x/ab545d.

[47] T. C, M. H and H. L, et al. A Novel Trapezoid-Type Stick–Slip Piezoelectric Linear Actuator Using Right Circular Flexure Hinge Mechanism. Ieee Transactions On Industrial Electronics 2017; 64: 5545-5552. DOI: 10.1109/TIE.2017.2677318.

[48] X. L, Q. G and Y. L, et al. A Linear Piezoelectric Stick-Slip Actuator via Triangular Displacement Amplification Mechanism. IEEE Access 2020; 8: 6515-6522. DOI: 10.1109/ACCESS.2019.2963680.

[49] J. Y, J. C and Y. H, et al. An Umbrella-Shaped Linear Piezoelectric Actuator Based on Stick-Slip Motion Principle. IEEE Access 2019; 7: 157724-157729. DOI: 10.1109/ACCESS.2019.2947449.

[50] Lu X, Gao Q and Gao Q, et al. Design, modeling, and performance of a bidirectional stick-slip piezoelectric actuator with coupled asymmetrical flexure hinge mechanisms. Journal of Intelligent Material Systems and Structures 2020; 31: 1961-1972. DOI: 10.1177/1045389X20942325.

[51] J. L, H. H and H. Z. A Piezoelectric-Driven Linear Actuator by Means of Coupling Motion. Ieee Transactions On Industrial Electronics 2018; 65: 2458-2466. DOI: 10.1109/TIE.2017.2740828.

[52] Y. Z, Y. P and Z. S, et al. A Novel Stick–Slip Piezoelectric Actuator Based on a Triangular Compliant Driving Mechanism. Ieee Transactions On Industrial Electronics 2019; 66: 5374-5382. DOI: 10.1109/TIE.2018.2868274.

[53] Zhang Y, Wang M and Cheng Y, et al. A stick-slip/inchworm hybrid rotary piezo motor based on a symmetric triangular driving mechanism. Applied Physics Letters 2019; 115: 131904. DOI: 10.1063/1.5119000.

[54] Zhang Y, Wang M and Fan Y, et al. Improving load capacity of stick-slip actuators in both driving directions via a shared driving foot. Smart Materials and Structures 2019; 28: 65004. DOI: 10.1088/1361-665x/ab1428.

[55] Zhang X, Yu Y and Gao Q, et al. A stick-slip linear piezoelectric actuator with mode conversion flexible hinge driven by symmetrical waveform. Smart Materials and Structures 2020; 29: 55035. DOI: 10.1088/1361-665x/ab7f42.

[56] Wang Y, Xu Z and Huang H. A novel stick-slip piezoelectric rotary actuator designed by employing

a centrosymmetric flexure hinge mechanism. Smart Materials and Structures 2020; 29: 125006. DOI: 10.1088/1361-665x/abb98c.

[57] Wang J, Qin F and Li L, et al. A linear piezoelectric actuator with high flexibility flexible mechanism designed by the bidirectional parasitic motion principle. Review of Scientific Instruments 2020; 91: 45005. DOI: 10.1063/5.0003382.

[58] Li X, Wang X and Sun W, et al. A compact 2-DOF piezo-driven positioning stage designed by using the parasitic motion of flexure hinge mechanism. Smart Materials and Structures 2019; 29: 15022. DOI: 10.1088/1361-665x/ab547f.

[59] Wang J, Huang H and Zhang S, et al. Development and analysis of a stick-slip rotary piezoelectric positioner achieving high velocity with compact structure. Mechanical Systems and Signal Processing 2020; 145: 106895. DOI: https://doi.org/10.1016/j.ymssp.2020.106895.

[60] Huang H, Li J and Zhao H, et al. On the correlation between the structure and one stepping characteristic of a piezo-driven rotary actuator. Microsystem Technologies 2016; 22: 2821-2827. DOI: 10.1007/s00542-015-2670-9.

[61] Wang J, Huang H and Wang Z, et al. Development and analysis of a dynamic model for parasitic motion principle piezoelectric actuator. Mechanical Systems and Signal Processing 2021; 147: 107079. DOI: https://doi.org/10.1016/j.ymssp.2020.107079.

[62] Fan H, Tang J and Li T, et al. Active suppression of the backward motion in a parasitic motion principle (PMP) piezoelectric actuator. Smart Materials and Structures 2019; 28: 125006. DOI: 10.1088/1361-665x/ab4e07.

[63] Tang J, Fan H and Liu J, et al. Suppressing the backward motion of a stick–slip piezoelectric actuator by means of the sequential control method (SCM). Mechanical Systems and Signal Processing 2020; 143: 106855. DOI: https://doi.org/10.1016/j.ymssp.2020.106855.

[64] Wang X, Huang H and Fan H, et al. Evolution of one-stepping characteristics of a stick-slip piezoelectric actuator under various initial gaps. Sensors and Actuators A: Physical 2019; 295: 348-356. DOI: https://doi.org/10.1016/j.sna.2019.06.023.

[65] Yang Z, Zhou X and Huang H, et al. A new motion mode of a parasitic motion principle (PMP) piezoelectric actuator by preloading the flexible hinge mechanism. Sensors and Actuators A: Physical 2019; 295: 396-404. DOI: https://doi.org/10.1016/j.sna.2019.06.025.

[66] Xu Z, Huang H and Dong J. A stick-slip piezoelectric actuator with measurable contact force. Mechanical Systems and Signal Processing 2020; 144: 106881. DOI: https://doi.org/10.1016/j.ymssp.2020.106881.

Chapter 3

Principle, Design and Future of Inchworm Type Piezoelectric Actuators

Jianping Li, Jianming Wen, Yili Hu, Zhonghua Zhang, Lidong He and Nen Wan

Abstract

The inchworm type piezoelectric actuator is one novel actuator to ensure a large working stroke with high resolution which has attracted the continuous attentions from researchers all over the world. In this study, the motion principle of the inchworm type piezoelectric is discussed: the "walker" pattern, the "pusher" pattern and hybrid "walker-pusher" pattern. The classification (linear, rotary and multi-DOF) and development are introduced in details, some significant researches are illustrated. Finally, the future direction of inchworm type piezoelectric actuators is pointed out according the development of inchworm type piezoelectric actuators. This study shows the clear principle, design and future of inchworm type piezoelectric actuators which is meaningful for the development of piezoelectric actuators.

Keywords: piezoelectric actuator, inchworm, large stroke, high resolution

1. Introduction

Up to now, researchers have developed a variety of micro/Nano driving and positioning platforms based on piezoelectric materials. Among the developed piezoelectric platforms, many of them have been applied for biological cell manipulation, atomic manipulation, micro/nano indentation, aerial photography and other systems with great application results [1, 2]. However, the working stroke of piezoelectric components is quite small, often only a few micrometers or tens of micrometers, which seriously limits the further application of piezoelectric actuators. Therefore, many researchers have done a lot of work to overcome this shortcoming of piezoelectric components, so as to expand the application field of piezoelectric actuators [3]. The inchworm type piezoelectric actuator is one kind of the developed new piezoelectric actuators which is able to ensure a large working stroke and achieve nano-scale accuracy at the same time. It has a wide application demand in the fields which have strict requirements on output accuracy, space size and antielectronmagnetic interference. The study on inchworm piezoelectric actuators has become a hot spot in the application and research field of piezoelectric actuators in recent years [4, 5].

IntechOpen

2. Motion principle

The inchworm type piezoelectric actuator mimics the motion principle of the real inchworm in nature, as is illustrated in **Figure 1(a)** [6]. Sometimes, it is called one kind of novel bionic actuators. It is found that the natural inchworm moves smoothly by stepping motion form. With the help of the stepping motion form and the piezoelectric technology, large working stroke is easy to be achieved by inchworm actuators through the alternating motion of driving units and clamping units. At the same time, compared with other piezoelectric actuators, the use of clamping unit brings larger output force. The inchworm type piezoelectric actuator usually consists of one driving unit and two clamping units. According to the difference of motion modes, inchworm type piezoelectric actuators could be split into three motion patterns: the "walker" pattern, the "pusher" pattern and hybrid "walker-pusher" pattern.

Figure 1(b) shows the motion principle of the "walker" pattern piezoelectric actuator, which is essentially similar to the walking mode of the real inchworm in nature. The "walker" mechanism obtains a large working stroke by repeating the following six steps: (1) in the original position, all piezoelectric elements in the driving unit and clamping units do not work, so there is a gap between the clamping device and the base guider; (2) the piezoelectric element in clamping unit 1 obtains the power, and then clamping unit 1 holds the base guide tightly; (3) the driving unit is extending while the piezoelectric element inside it obtains the power; (4) the clamping unit 2 obtains the power to tightly fix the base guider and the clamping unit 1; (5) the clamping unit 1 loses power to release the base guider; (6) the driving unit returns to its original length. Finally, the clamping unit 2 is de energized to release the base guider in the same original position as in step (1). By repeating these six steps, a large working stroke is achieved gradually [7].

The "push" pattern piezoelectric actuator also needs six steps to obtain a step motion, as shown in **Figure 1(c)**: (1) in the original position, all of the driving and clamping units have no power to fix the slider; (2) the clamping unit 2 obtains the power to hold the slider tightly; (3) the driving unit gets power to push the clamping unit 2, and since the slider is hold by clamping unit 2 tightly, it will move forward for small moving distance; (4) the clamping unit 1 holds the slider; (5) the clamping unit 2 is powered off to release the slider; (6) the driving unit returns to the original length when it loses power. At last, the clamping unit 1 releases the slider to the same condition as in the original position [8].

Hybrid "Walker-pusher" pattern piezoelectric actuator is a hybrid of "Walker" and "pusher" modes. The difference is that the driving unit is inserted into the sliding block, and the clamping unit is assembled in the base in the "Walker-pusher" mode [9].

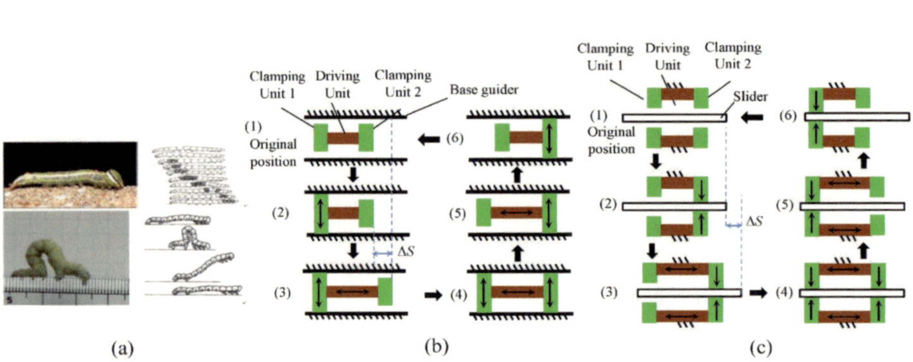

Figure 1.
Motion principles: (a) real inchworm; (b) "walker" pattern; (c) "pusher" pattern.

3. Classification and development

Inchworm type piezoelectric actuator has been widely concerned by researchers because it is able to ensure long working stroke, high precision and large output at the same time. Many inchworm driving devices have been developed. According to the different motion forms, they could be divided into the linear actuator, the rotary actuator and the multi-DOF actuator.

3.1 Linear inchworm actuator

As early as in 1964, Stibitz developed the first "pusher" inchworm actuator with magnetostrictive elements to generate driving force, so as to solve the positioning problem of machining tools [10]. Three magnetostrictive elements are utilized as driving and clamping units to generate large stroke linear stepping motion. However, limited by the technical conditions, its performance is not very high, but the driving principle of the inchworm movement provides a new space for the research of precision positioning technology. After that, Hsu et al. proposed the inchworm piezoelectric actuator for the first time [11]. As shown in **Figure 2(b)**, the piezoelectric element is applied to convert the electrical signal into mechanical motion, and two unidirectional clamps are combined to accumulate the movement of the electric element. A piezoelectric tube is inserted into the slider so that it is in the mixed "walker-pusher" mode. In 1968, Brisbane invented the first "walker" type inchworm piezoelectric actuator [12]. Two piezoelectric disks and one piezoelectric tube are assembled inside the slider, which makes it possible to realize the linear motion of the slider by walking which is illustrated in **Figure 2(c)**.

However, due to the immaturity of piezoelectric materials at that time, the development of inchworm type piezoelectric actuators was hindered. For many years, even though there were still some researches on inchworm type piezoelectric actuators, most of them only focused on the theoretical research. Until the end of the 1980s, commercial piezoelectric elements were able to provide an output force of up to several thousand newtons, and the driving voltage dropped from 1000 V to 200 V. All of these provided great opportunities for the further development of piezoelectric actuators. After that many researchers have focused on the development of inchworm type piezoelectric actuators. By using three packaged piezoelectric stacks forming a U-shaped structure, Chen et al. proposed a "pusher" pattern inchworm type piezoelectric actuator [13]. The experimental results show that the maximum driving force is 13.2 N and the maximum speed is 47.6 μm/s. With the help of adding an integrated heterodyne interferometer as feedback device in the servo control system, an inchworm type piezoelectric actuator with fast response is developed by Moon et al. [14]. Based on the fast response characteristics of the servo control system, it can move to the target position quickly and reduce the hysteresis of the piezoelectric actuator.

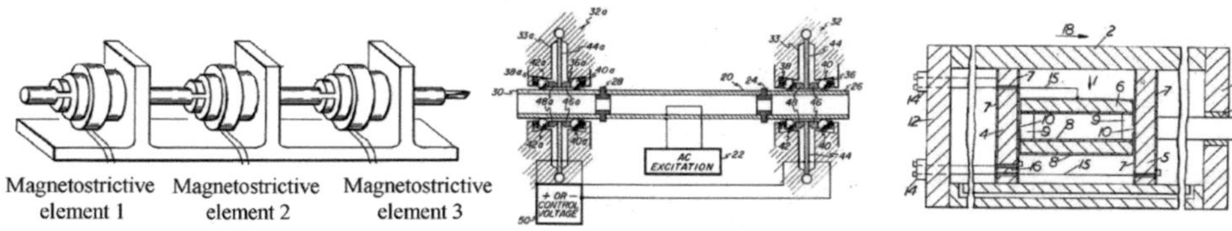

Figure 2.
Inchworm actuators: (a) the first inchworm actuator [10]; (b) the first inchworm type piezoelectric actuator [11]; (c) the first "walker" pattern inchworm piezoelectric actuator [12].

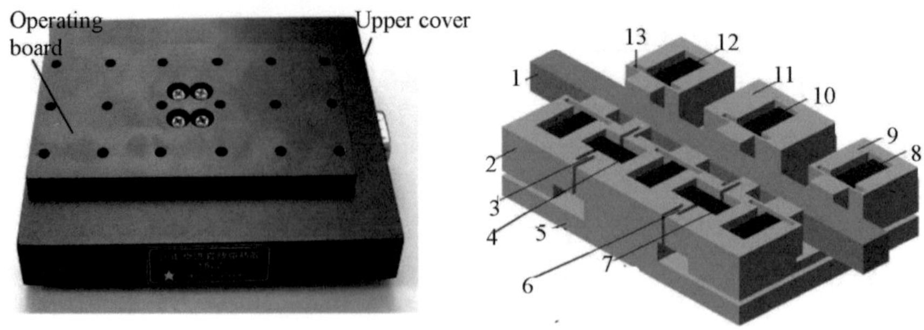

Figure 3.
Inchworm actuator developed by Yang et al. [17].

However, an important problem is that the extension length of piezoelectric elements is very small, which brings trouble to the clamping unit to clamp the slider tightly. Therefore, many literatures are focusing on different methods to improve the clamping unit. As a typical compliant mechanism, the flexure hinge mechanism has been widely applied in the design of piezoelectric actuators to expand the elongation of piezoelectric elements due to its advantages of fast response, no friction and easy manufacturing. In 1988, Fujimoto firstly proposed an inchworm type piezoelectric actuator with flexible hinge [15]. This "walker" type piezoelectric actuator adopts C-shaped lever type flexible hinge on both clamping units to increase the clamping force, and it has great practical value for the real application of inchworm piezoelectric actuators. The magnification could be adjusted by changing the position of the pivot point. Kim constructed an inchworm platform with an amplification stage, and it utilized the flexure hinge as a lever mechanism to obtain a magnification of 8.4 at a leverage ratio of 3.6 [16].

The research team of Jilin University and Zhejiang Normal University has carried out systematic research on the development of inchworm piezoelectric actuators. After years of experience, it has developed series of Inchworm piezoelectric actuators, and has achieved a series of remarkable research results. For example, Yang et al. proposed a novel linear piezoelectric actuator [17] (**Figure 3**). The proposed actuator adopts the principle of "pusher" motion pattern, and realizes the passive linear motion of the slider with the help of clamping and driving units. Based on the analysis of the working principle and the mechanical structure of the actuator, a linear driving mathematical model with the piezoelectric stack as the driving element is established, and its structure is analyzed by finite element method (FEM). The proposed inchworm piezoelectric actuator adopts the principle of bidirectional thrust, and realizes the consistency of driving characteristics in the process of forward and reverse directions. Experimental results show that the novel inchworm actuator has the characteristics of firm clamping, high frequency (100 Hz), high step speed (30 mm/min), large stroke (> 10 mm), high resolution (0.05 μm) and large driving force (100 N), which greatly improves the driving performance of the inchworm piezoelectric actuator. It has a broad further application in precision motion, micromanipulation, optical engineering, and precise positioning and so on.

3.2 Rotary inchworm actuator

Besides the inchworm piezoelectric actuators to achieve linear motion, some inchworm actuators which could obtain rotary motion have been developed by researchers. Kim et al. developed a new type of inchworm piezoelectric actuator that uses a combination of flexure hinge and piezoelectric drive technology to

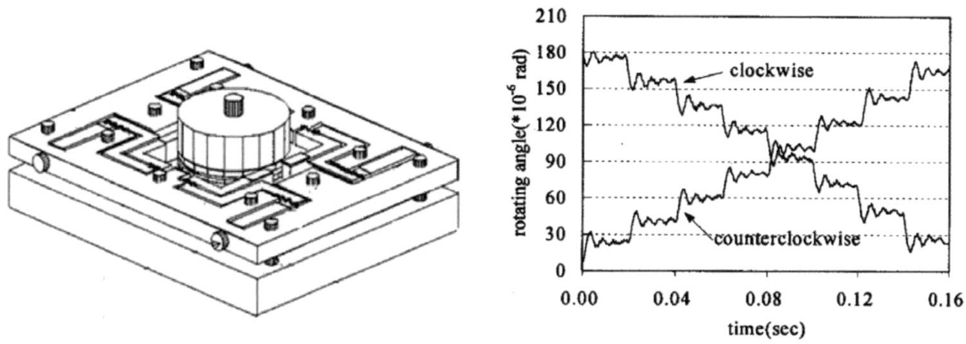

Figure 4.
Inchworm type piezoelectric actuators with flexible belts by Kim et al. [18].

achieve rotational movement [18] (**Figure 4**). The device pioneered the use of linear output piezoelectric stacks to achieve an inchworm-shaped rotary motion, which has extremely high research significance. The device realizes the movement of the flexible hinge by controlling the power-on sequence of the four piezoelectric stacks, thereby driving the belt wound on the rotating shaft to drive the rotating shaft to rotate. The test results show that the resolution of the rotary drive device can reach 2.36 μrad, which is greatly improved compared to the previous rotary drive device.

In view of the shortcomings of the existing inchworm actuators, Li et al. firstly designed an inchworm type piezoelectric actuator based on multi-layer torsional flexure hinges, which is able realize the rotary motion with large working stroke and high precision [4]. The developed actuator utilizes the piezoelectric stack to push the thin-walled flexure hinge structure to carry out relevant clamping. By controlling the working sequence of the clamping units in the first and second layers of the stator, the precise rotary motion around the fixed shaft is realized step by step. Its structure is divided into two main parts: rotor and stator. According to the function, it could be divided into the driving unit, the clamping unit and the preloading unit. The proposed device uses high-precision piezoelectric stack to push the thin-walled flexure hinge structure for relevant clamping. By controlling the clamping sequence of the piezoelectric clamping units in the first and second layers of the stator, the step-by-step ultra-precision rotary motion around the rotating shaft is realized. The stator is packaged with two layers of the self-centering piezoelectric clamping unit, rotary driving unit and preloading unit; the rotor is a variable interface rotating shaft, which can drive different objects by changing the connection style of the interface. The clamping unit is composed of the piezoelectric stack encapsulated in the stator and the self-centering flexure hinge. The preloading unit is utilized to pre tighten the clamping piezoelectric stack, and the clamping pressure is adjusted by adjusting the screw in length to control the engaging wedge block. The driving unit is composed of the driving piezoelectric stack, the driving indenting block and the corresponding parts of stator, which is used to apply rotating torque to the first layer of stator. The maximum diameter of stator is 80 mm and the diameter of rotor is 20 mm. This proposed inchworm type piezoelectric actuator could achieve stable stepping rotation output. The size of the driving voltage will affect the single-step rotation angle of the rotor: as the driving voltage increases, the rotation angle of the rotor also increases; when the driving voltage is less than 20 V, the rotor cannot work stably, so the minimum step angle of the rotor is 4.95 μrad. In the case that the driving voltage is 100 V, the maximum step angle of the rotor is 216.7 μrad. The maximum speed of the rotor is 6508.5 μrad/s, and the driving frequency is 30 Hz. The designed inchworm type piezoelectric actuator has a maximum output torque of 93.1 N·mm. **Figure 5** shows that the driving voltage and clamping voltage are maintained at 100 V, and when the driving frequency is 1 Hz, after the rotor rotates 20 steps in the forward and reverse directions, the forward and backward error of the rotor is 0.76 μm. The total error of 20 steps is 38 μrad, so the step angle error of the inchworm type piezoelectric rotary actuator designed in this paper is 1.9 μrad.

The disadvantage of the inchworm type piezoelectric actuator is that the structure is relatively complicated. The traditional inchworm type actuator needs to use at least two clamping units and one drive unit. In this way, multiple timing controls will cause the program to be complicated, which makes the inchworm piezoelectric actuator more complicated. The application has brought unfavorable effects. Based on the above work, a simplified inchworm type piezoelectric rotary actuator was designed and manufactured by Li et al., which uses a triangular lever flexure hinge to complete the clamping and driving actions at the same time [19].

By using the triangular lever flexure hinge, one driving unit and one clamping unit could be utilized to realize stepping rotation of the rotor. Its stator is simplified from a two-layer structure to a single-layer structure, which reduces the overall height; the control adopts two-channel voltage control, which reduces the output of one clamp voltage. **Figure 6** shows the overall structure of the simplified inchworm piezoelectric actuator, which mainly includes a stator, a rotor, four drive piezoelectric stacks, two clamp piezoelectric stacks and six pre-tightening screws. The stator material is 65Mn, and the drive hinge and clamp hinge are processed by wire cutting. The rotor diameter is 20 mm. The pre-tightening screws are used to adjust the pre-tightening force of the clamping piezoelectric stack and the driving piezoelectric stack. It is seen from **Figure 6** that there is a small "jump" in the middle of each step, which is caused by the impact of the clamping unit on the rotor. When the driving voltage is 100 V and the driving frequency is 1 Hz, the maximum output torque of the designed simplified inchworm piezoelectric actuator is 19.6 N·mm. When the output load is greater than 19.6 N·mm, the rotor cannot run stably. When the driving voltage signal increases from 20 V to 100 V, the rotor step angle also increases, which coincides with the approximately proportional relationship between the output displacement of the piezoelectric stack and the driving voltage. The maximum step angle occurs when the drive voltage is 100 V and the drive frequency is fixed at 1 Hz, and the maximum step angle is 1360 μrad. When the drive voltage is less than 20 V, the simplified inchworm piezoelectric actuator cannot operate stably, so its operating resolution is 25 μrad. Contrary to the above, when the drive frequency is increased from 0 Hz to 200 Hz, the rotor step angle decreases rapidly. After 200 Hz, the rotor step angle stabilizes near a small value.

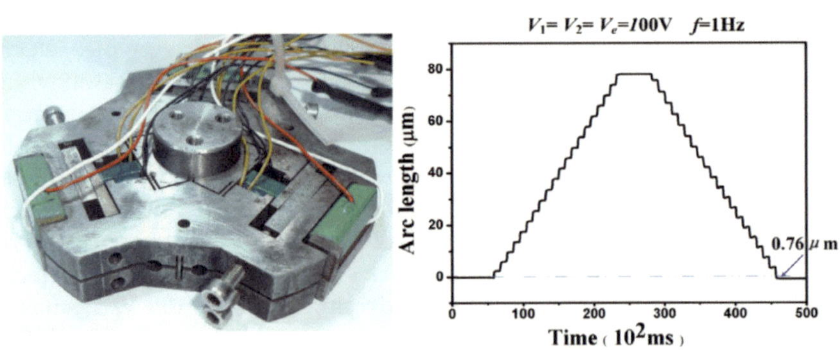

Figure 5.
Rotary inchworm type piezoelectric actuators by Li et al. [4].

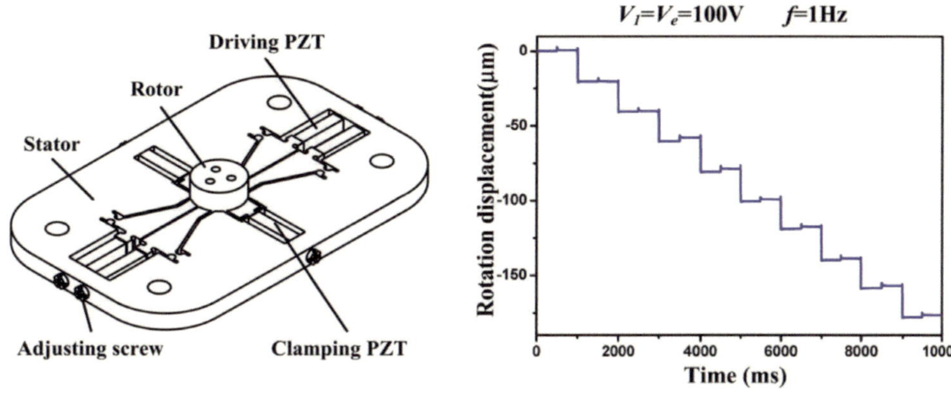

Figure 6.
Simplified inchworm piezoelectric actuator by Li et al. [19].

3.3 Multi-DOF inchworm actuator

How to obtain multi-DOF motion within a compact size is always the pursuing interest for researchers of the actuator field. Generally, same single-DOF actuators are assembled in series to achieve the so called multi-DOF motion, which brings the large structure size and assemble problems. With the help of integral flexure structure, Li et al. firstly proposed the 2-DOF inchworm piezoelectric actuator which could achieve both rotary and linear motions with a compact size, as is shown in **Figure** 7 [20]. The structure of the proposed 2-DOF actuator is composed of a stator and a slider. The stator and slider are subdivided into upper, middle and lower layers. Four right-angle flexure hinges acting as torsion springs are used to overlap the upper and middle layers of the stator. The linear displacement of the positioning platform relies on four flexure hinges to connect the middle and lower layers of the stator. Moreover, according to the characteristics of PZTs that can be driven by linear motion and rotational motion, four linear driving PZTs and one rotary driving PZT are respectively arranged on the upper and lower layers of the stator. As for the slider, each layer is fixed with a single clamping PZT. Using 65Mn as the material of the stator and slider to obtain higher elasticity, the device needs to be vacuum heat treated.

The positioning platform can realize linear movement and rotational movement according to different numbers of piezoelectric ceramics, placement

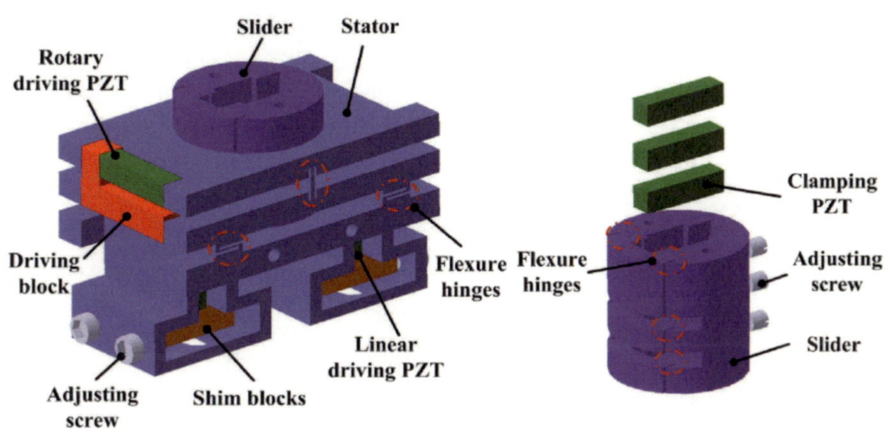

Figure 7.
Graphic model of the 2-DOF inchworm piezoelectric actuator [20].

positions and flexible hinges. For the rotary motion, the proposed actuator operates stably under a driving voltage of 100 V to 6 V. In the case that the driving voltage is reduced from 100 V to 6 V, the rotation angle of 10 steps decreases. This result may be that the degree of PZT expansion is directly proportional to the input voltage. In addition, with the lowering of the driving voltage, the amplitude of the first-order oscillation decreases from 28.20μrad to 3.75μrad. During the down-regulation process, it is found that the step displacement of the platform is shortened and the fluctuation amplitude is larger. The platform cannot work stably when the driving voltage is lower than 6 V. According to the total rotation angle of 4.52μrad, 20 steps, the minimum step angle is 0.23μrad. It indicates that this inchworm positioning platform has good performance under constant driving frequency and driving voltage. Under the condition of controlling the driving frequency, the speed increases with the increase of the driving voltage. When f = 21 Hz, the speed reaches the peak value. When f = 20 Hz and U = 100 V, the maximum speed is 3521.70 μrad/s. However, when the frequency is higher than 21 Hz, the mechanical structure of the drive cannot normally respond to the electrical signal. The energy conversion method can be explained as that the structure cannot convert all the electrical energy into mechanical energy due to the high frequency, and there will be a certain energy loss.

For the linear motion, the designed inchworm actuator works continuously under a constant driving voltage of 10 V to 100 V. Under the driving voltage U = 100 V, the total displacement of the actuator in 10 steps is 82.30 μm, and the available single-step displacement is 8.23 μm. In the case that the input voltage is lower than 10 V, the actuator cannot work normally. According to the total rotation angle of 3.05μrad, 20 steps, the minimum step angle is 0.15 μm. The speed characteristics of the linear motion of the actuator under the clamp voltage have been mentioned. According to the experimental data, as the frequency increases, the speed increases. When the frequency is greater than 26 Hz, the speed gradually decreases. When f = 26 Hz and U = 100 V, the maximum speed is 105.31 μm/s.

3.4 Comparison

As shown in **Table 1**, three types of inchworm actuators all obtain large output force/torque and stroke, high resolution. Previous studies indicate that all types are able to realize the output force/torque of several to dozen newton/newton metre. The resolution scales of them all attain micrometer/microradian and based on its working principle, repeating the displacement output under the periodic signal, their stroke are all very large. Linear inchworm actuator is able to attain a high speed of 30 mm/min and rotary inchworm actuator achieves a high speed of 6508.5 μrad/s while Multi-DOF inchworm actuator is slower. To achieve the aim of multi-DOF, the structure of Multi-DOF inchworm actuator is also more complicated with a slower response.

Type	Structure	Motion type	Output speed	Output force/torque	Stroke
Linear	Simple	Linear	Large	Large	Large
Rotary	Medium	Rotary	Large	Large	Large
Multi-DOF	Complicated	Multi-DOF	Medium	Large	Large

Table 1.
Characteristics comparison of different inchworm actuators.

3.5 Applications

Over the past years, the inchworm actuator has been widely applied in some commercial areas. High resolution is one of the most significant advantages of the inchworm actuator. Therefore, ultra-precision manufacturing technology, precision focusing system and micro-robot obtains wide range use of inchworm actuator as their actuation sources [21–22]. When coupled with large output force/torque, the inchworm actuator is also widely used in medical engineering areas like drug delivery, cell manipulation, lab on a chip [23–24]. Compared with other piezoelectric actuators, the advantage of long stroke also employs inchworm actuator in precision position platform [25].

4. Future direction

4.1 Simplified structure

One of the significant shortcomings of inchworm type piezoelectric is the complex structure which brings trouble for the manufacture and control. **Figure 8** shows the structure of the proposed simplified piezoelectric actuator based on the parasitic movement of the flexure mechanism by Li et al. [26]. With the help of the parasitic movement of the flexure mechanism, only two piezoelectric elements are needed. It is mainly composed of the base, the slider, piezo-stack 1, piezo-stack 2, flexure mechanism 1, flexure mechanism 2, two wedge blocks, four micrometer knobs and eight screws. Piezo-stack 1 (AE0505D16, 5 × 5 × 20 mm, NEC/TOKIN CORPORATION) is inserted into the flexure mechanism 1 through the wedge block to push the linear slider. The assembly process of the piezoelectric stack 2 and the flexure mechanism 2 are the same. The high-precision four-micron knob (M6 from SHSIWI) is utilized to adjust the preloading force between the flexure mechanism and the slider. The slider is a commercial linear guide with high linearity produced by THK. The flexure mechanism is made of aluminum alloy AL7075 manufactured by WEDM. Screws are applied to stably assemble all components on the base. The overall size of the proposed stepping piezoelectric actuator is 100 mm × 60 mm × 18 mm.

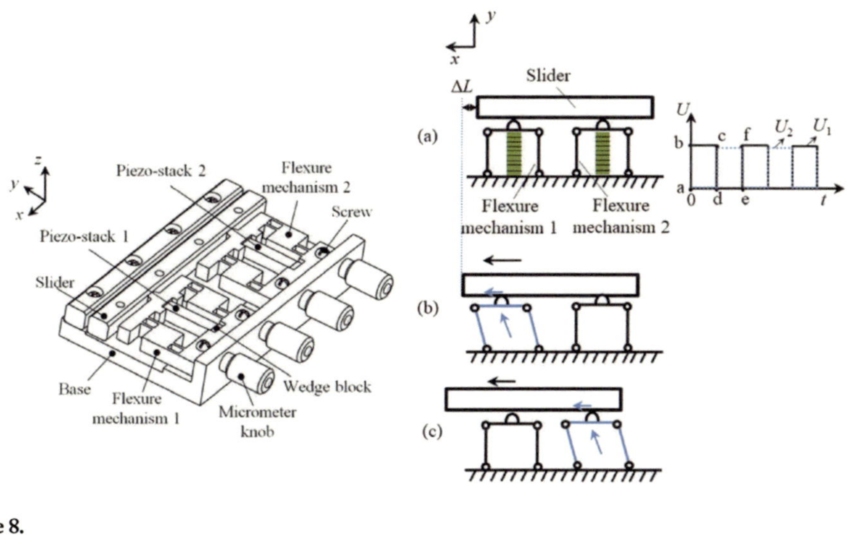

Figure 8.
Structure of the proposed simplified piezoelectric actuator and motion principle [26].

Two piezoelectric "legs" are required to alternately drive the slider, and this is why they are sometimes called "walking" type piezoelectric actuators. In addition, for traditional "walking" type piezoelectric actuator, in each piezoelectric "leg", at least two piezoelectric elements are required (one for flexure movement and one for longitudinal movement). The movement principle of the stepping piezoelectric actuator is the "circular movement" of the piezoelectric "legs". In short, each piezoelectric "leg" should achieve two movements in x and y directions.

In the proposed study by Li et al., the parasitic movement of the flexure mechanism is applied to simplify the entire system. Generally, the piezo-stack could only achieve the one motion in its longitudinal direction. Whereas, as shown in **Figure 8(b)**, with the aid of the asymmetrical flexure mechanism, the piezo-stack will generate an oblique upward force, which causes the motion displacement in both x and y directions. The parasitic movement L_x in x direction is used to drive the linear movement of the slider. However, only one flexure mechanism cannot achieve walking motion, and at least two flexure mechanisms ("legs") are required. In addition, during the movement, the input square wave voltages U_1 and U_2 have the same magnitude but different phases. The experimental results display that the application of the parasitic motion of the flexure mechanism is able to simplify the inchworm type piezoelectric actuator. The stepping motion of the proposed actuator requires only two piezoelectric elements and two input signals. Additionally, performance of the proposed simplified piezoelectric actuator (stepping performance, speed performance, and load performance) has a certain relationship with the input voltage and frequency. Under the conditions of U = 100 V and f = 1 Hz, the maximum step displacement ΔL = 1.75 μm. Under the condition of U = 30 V and f = 1 Hz, the minimum step displacement ΔL = 0.18 μm. When U = 100 V, f = 20 Hz, the maximum movement speed V_s = 39.78 μm. This study verifies the feasibility of design and simplification of inchworm type piezoelectric actuators with parasitic motion of flexure mechanisms, and provides a new idea for the research of piezoelectric actuators. Potential applications in optical engineering and cellular operating systems require more work.

4.2 Simplified control

For most of the inchworm type piezoelectric actuators, three input signals are necessary for one driving unit and two clamping units, which make the control system also complicated. In order to simplify the control system, Gao et al. proposed one novel piezoelectric inchworm actuator which uses a DC motor to drive the permanent magnet for alternate clamping, applies a laser beam sensor to detect the position of the permanent magnet and generates an excitation signal to drive the piezoelectric stack [27]. The actuator only needs a DC signal to drive and can adjust the frequency by changing the motor speed. The movement mechanism of the actuator is emphatically discussed, and the influence of the permanent magnet structure on the clamp is studied. The flexibility matrix method and COMSOL finite element software are used to simulate and analyze the flexure hinge. The driving signal for the piezoelectric stack is generated by self-sensing and automatically adapts to the frequency change, which simplifies the control signal of the inchworm actuator. The use of the magnetic clamping unit solves the serious friction and wear problems of the current clamping method of piezoelectric inchworm actuators. In addition, the driving unit and clamping unit of the proposed piezoelectric inchworm actuator are tested experimentally. The experimental results confirm the feasibility of the proposed scheme and obtained relevant optimized structural parameters.

The overall structure of the proposed actuator, as shown in **Figure 9**, is mainly composed of a sensing unit, a driving unit and a clamping unit. As shown in **Figure 9(a)**, the clamping unit is mainly composed of a DC motor, a motor base, a permanent magnet after magnetization (RPM, red), a permanent magnet before magnetization (NRPM, blue), bearings and a bearing housing. As shown in **Figure 9(b)**, the sensing unit includes a cam, a laser beam sensor (OLS) and a bracket. The driving unit includes a flexible hinge mechanism with integrated piezoelectric stack (AE0505D16, NEC/TOKIN CORPORATION), a wedge-shaped adjusting mechanism (built-in a pair of wedges and a pre-tightening bolt) and a slider, as shown in **Figure 9(c)**. The designed slider can slide in the sliding groove of the flexible hinge mechanism. Two clamping modules and cams are fixed at the end of the output shaft of the DC motor, and each clamping module is assembled by a radially polarized permanent magnet RPM and a non-radially polarized permanent magnet NRPM. The piezoelectric stack is preloaded by the wedge-shaped adjusting mechanism and nested in the installation slot of the flexible hinge mechanism. The laser beam sensor is supported by two brackets and generates an excitation signal by detecting the position of the cam. In addition, the support block, the DC motor and the bearing are assembled on the base with eight bolts.

The proposed inchworm actuator by Gao et al. utilizes a DC motor to drive the permanent magnet for rotating to achieve alternate clamping. The actuator does not need to input the driving voltage signal of the piezoelectric stack. It only senses the position of the permanent magnet through the laser beam sensor, and generates an excitation signal to drive the piezoelectric stack to achieve precise linear displacement output. Its working principle is shown in **Figure 10**. Work performance of

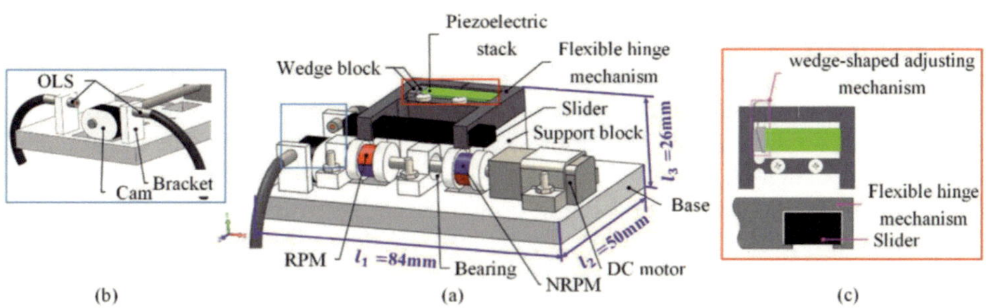

Figure 9.
Structure of the actuator by Gao et al.: (a) sensing unit; (b) sensing unit; (c) driving unit [27].

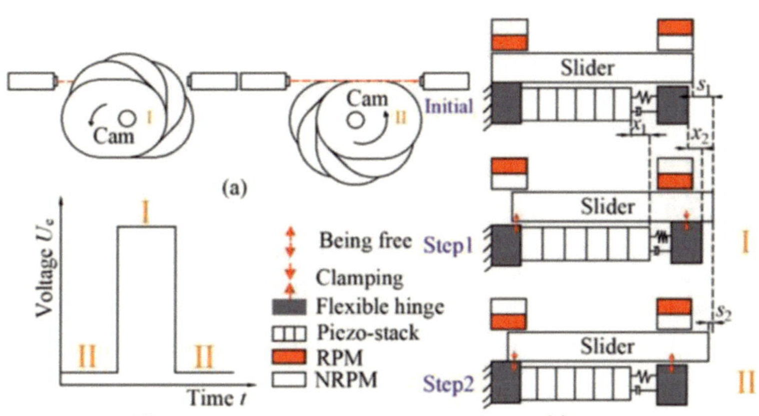

Figure 10.
Working principle of the inchworm piezoelectric actuator with simplified control system by Gao et al. [27].

(a) (b)

Figure 11.
Miniaturized inchworm type actuators: (a) by Risaku et al. [28]; (b) by Mehmet et al. [29].

the proposed actuator was studied carefully. For the important component of the driving unit, the "Z" type flexure hinge, the flexibility matrix method is used to perform theoretical calculations. The error between the simulation results and the theoretical calculation results is about 2.13%, indicating the accuracy of the calculation; for the magnetic clamping unit, when the clamping distance is 1 mm, the magnetic clamping unit has better clamping capability. The experimental results show that the actuator has a good linear displacement. When U_e = 150 V and f = 40 Hz, its maximum movement speed is 481.43 μm/s, and the maximum load is m = 950 g.

4.3 Other directions

The inchworm type piezoelectric drive device can not only obtain large output stroke, but also ensure high output accuracy and load-carrying capacity, which is favored by many scholars. The research of Inchworm piezoelectric driving device has its own characteristics at home and abroad, which provides a favorable technical basis for the development and application of piezoelectric precision drive technology. Besides the above future directions, the existing inchworm piezoelectric actuator is still in the stage of empirical design and test, lacking of relevant theoretical model guidance, and there are problems of empirical design and repeated attempts. Therefore, it is necessary to establish the dynamic model of the inchworm piezoelectric actuator to guide the design and research of the inchworm piezoelectric drive device. In addition, the miniaturization is always the hot point for piezoelectric actuators which could leads to the real application in many research and industrial fields.

Risaku et al. have developed a large stroke and high precision inchworm actuator [28] (**Figure 11**). With the combination of piezoelectric and electrostatic motion principles, the displacement accuracy of each step reaches tens of nanometers, which can be called ultra-high precision. The displacement accuracy is 59 nm/cycle, but the maximum travel distance is only 600 μm, which needs to be improved.

In order to solve the shortcomings that most inchworm type piezoelectric actuators require larger input voltage, Mehmet et al. took the lead in developing a new type of low voltage, largestroke, and large output inchworm actuator based on the micro-electromechanical systems (MEMS) [29]. It mainly applies the principle of electrostatic motion. Through the amplification of the flexible hinge, it achieves a total displacement of ±18 μm and an output force of ±30 μN at a low voltage of 7 V; a displacement of ±35 μm can be achieved at a voltage of 16 V, ±110 μN output force.

5. Conclusion

The inchworm movement is a high-precision driving method that imitates the movement form of the inchworm in nature to realize the stepping movement of itself or the holding object. Inchworm movement is a kind of stepping movement, which is different from other continuous movement. Its movement can be regarded as a combination of movement and stop in time, but it is also a continuous movement from the perspective of the overall effect of the movement. Inchworm motion can easily achieve large-stroke step-by-step linear motion. Although scholars in various countries have conducted a number of research work on inchworm-type piezoelectric driving devices, most of their research content is linear inchworm driving devices, which involve rotation. There are few reports on the inchworm actuator, and the existing inchworm-type piezoelectric actuator has complex structure control, lacks relevant theoretical model guidance, and has problems of empirical design and repeated attempts. In the future, there is still a lot of work to be solved for the inchworm piezoelectric actuator to promote the real practical use of the inchworm piezoelectric actuator.

Acknowledgements

This work is supported by the Natural Science Foundation of Zhejiang Province: LY19E050010, LY20E050009, LGF20E050001; General Research Projects of Zhejiang Provincial Department of Education: Y201943038; Zhejiang Provincial Key Research and Development Project of China: 2021C01181.ADDIN EN.REFLISTX.

Author details

Jianping Li*, Jianming Wen, Yili Hu, Zhonghua Zhang, Lidong He and Nen Wan
The Institute of Precision Machinery and Smart Structure, College of Engineering, Zhejiang Normal University, Jinhua, Zhejiang, China

*Address all correspondence to: lijp@zjnu.cn

IntechOpen

© 2021 The Author(s). Licensee IntechOpen. This chapter is distributed under the terms of the Creative Commons Attribution License (http://creativecommons.org/licenses/by/3.0), which permits unrestricted use, distribution, and reproduction in any medium, provided the original work is properly cited. CC BY

References

[1] M. W. Fairbairn, S. R. Moheimani, Fleming, A. J. Q Control of an Atomic Force Microscope Microcantilever: A Sensorless Approach, J. Microelectromech. S., 20(6) (2011) 1372-1381.

[2] Q. Li, L. Liu, X. Ma, et al, Development of Multi-target Acquisition, Pointing and Tracking System for Airborne Laser Communication, IEEE T. Ind. Inform. 15(3) (2018) 1720-1729.

[3] Y. Peng, X. Gu, J. Wang, H. Yu, A review of long range piezoelectric motors using frequency leveraged method, Sensor. Actuat. A: Phys. 235 (2015) 240-255.

[4] J. Li, H. Zhao, H. Qu et al, A piezoelectric-driven rotary actuator by means of inchworm motion, Sensor. Actuat. A: Phys. 194 (2013) 269-276.

[5] C. Moon, S. Lee, J. K. Chung, A new fast inchworm type actuator with the robust I/Q heterodyne interferometer feedback, Mechatronics 16(2) (2006) 105-10.

[6] Li J , Huang H , Morita T . Stepping Piezoelectric Actuators with Large Working Stroke for Nano-positioning Systems: A Review[J]. Sensors and Actuators A Physical, 2019, 292:39-51.

[7] B. Zhang, Z. Zhu, Developing a linear piezomotor with nanometer resolution and high stiffness, IEEE/ ASME Trans. Mechatron. 2(1) (1997) 22-29.

[8] D. Newton, E. Garcia, G.C. Horner, A linear piezoelectric motor, Smart Mater. Struct. 7(3) (1998) 295.

[9] S. Hua, G. Liu, X. Wang, Y. Wang, J. Li, A Piezoelectric Inchworm Actuator with Bidirectional Thrust Force, Sensors & Transducers, 171(5) (2014) 169.

[10] G. R. Stibitz, Incremental feed mechanisms, US Patent US3138749 (1964).

[11] S. K. Hsu, A. Arbor, A. Blatter, Transducer, US Patent US3292019 (1966).

[12] A.D. Brisbane, Position control device, US Patent US 3377489 (1968).

[13] W. Chen, Y. Liu, Y. Liu et al, Design and experimental evaluation of a novel stepping linear piezoelectric actuator, Sensor. Actuat. A: Phys. 276 (2018) 259-266.

[14] C. Moon, S. Lee, J. K. Chung, A new fast inchworm type actuator with the robust I/Q heterodyne interferometer feedback, Mechatronics 16(2) (2006) 105-10.

[15] T. Fujimoto, Linear motor driving device, US Patent US4736131 (1988).

[16] Y. Kim, S. Choi, J. Park et al, The characteristics of variable speed inchworm stage using lever mechanism by different materials, J. Nanosci. Nanotechno. 8(11) (2008) 5696.

[17] Yang Shuchen, Liu Jianfang, Yang Zhigang, et al. Research on external driven bidirectional thrust piezoelectric stepping precision actuator [J]. Optical precision engineering, 2006. (In Chinese)

[18] S. Kim, S. Kim, Precise rotary motor by inchworm motion using dual wrap belts, Rev. Sci. Instrum. 70(5) (1999) 2546-50.

[19] Li J , Zhao H , Shao M , et al. Design and experiment performances of an inchworm type rotary actuator[J]. Review of Scientific Instruments, 2014, 85(8):256.

[20] J. Li, H. Zhao, X. Qu et al, Development of a compact 2-DOF

precision piezoelectric positioning platform based on inchworm principle, Sensor. Actuat. A: Phys. 222 (2015) 87-95.

[21] Mohith S , Upadhya A , Navin K P , et al. Recent trends in piezoelectric actuators for precision motion and their applications: a review[J]. Smart. Mater. Struct. 2021, 30(013002).

[22] Wang L , Chen W , Liu J , et al. A review of recent studies on non-resonant piezoelectric actuators. Mech. Syst. Signal. Pr, 2019, 133(106254).

[23] Wang S , Rong W , Wang L , Xie et al. A survey of piezoelectric actuators with long working stroke in recent years: classifications, principles, connections and distinctions. Mech. Syst. Signal. 2019, 123: 591-605.

[24] Wang P , Xu Q . Design and testing of a flexure-based constant-force stage for biological cell micromanipulation. IEEE Transactions on Automation Science and Engineering, 2017 99: 1-13.

[25] Shubao S , Siyang S , Minglong X , Wenjian J . Mechanically reconfigurable reflector for future smart space antenna application. Smart. Mater. Struct. 2018 (accepted). DOI: 10.1088/1361-665X/aad480.

[26] Wen J , Wan N , Wang R , et al. A Novel Linear Walking Type Piezoelectric Actuator Based on the Parasitic Motion of Flexure Mechanisms[J]. IEEE Access, 2019:1-1.

[27] Gao Y , Wen J , Ma J , et al. A self-adapting linear inchworm piezoelectric actuator based on a permanent magnets clamping structure[J]. Mechanical systems and signal processing, 2019, 132(Oct.1):429-440.

[28] Risaku, Toda, Eui-Hyeok, et al. A normally latched, large-stroke, inchworm microactuator[J]. Journal of Micromechanics and Microengineering, 2007, 17(8).

[29] Erismis M A , Neves H P , Puers R, et al. Saw-tooth vernier ratchets for electrostatic inchworm actuators[J]. Sensors & Actuators A Physical, 2009, 156(1):66-71.

Chapter 4

The Asymmetric Flexure Hinge Structures and the Hybrid Excitation Methods for Piezoelectric Stick-Slip Actuators

Tinghai Cheng, Xiaosong Zhang, Xiaohui Lu, Hengyu Li, Qi Gao and Guangda Qiao

Abstract

Piezoelectric stick–slip actuators have become viable candidates for precise positioning and precise metering due to simple structure and long stroke. To improve the performances of the piezoelectric stick–slip actuators, our team deeply studies the actuators from both structural designs and driving methods. In terms of structural designs, the trapezoid-type, asymmetrical flexure hinges and mode conversion piezoelectric stick–slip actuators are proposed to improve the velocity and load based on the asymmetric structure; besides, a piezoelectric stick–slip actuator with a coupled asymmetrical flexure hinge mechanism is also developed to achieve the bidirectional motion. In terms of driving methods, a non-resonant mode smooth driving method (SDM) based on ultrasonic friction reduction is first proposed to restrain the backward motion during the rapid contraction stage. Then, a resonant mode SDM is further developed to improve the output performance of the piezoelectric stick–slip actuator. On this basis, the low voltage and symmetry of the SDM are also discussed. Finally, the direction-guidance hybrid method (DGHM) excitation method is presented to achieve superior performance, especially for high speed.

Keywords: piezoelectric stick–slip actuator, structural design, driving method, flexure hinge mechanism, ultrasonic friction reduction

1. Introduction

Piezoelectric stick–slip actuators have obtained considerable significance and have become the focus area of research for camera focusing mechanisms, cell phones, scanning probe micro-scopes, zoom lens systems and blue-ray devices, because of compact structure, low cost, theoretically unlimited displacement and convenient control [1–3]. Based on the configuration designs, the piezoelectric stick–slip actuators are divided into parallel type [4–6] and non-parallel type [7–9], as illustrated in **Figure 1**.

The parallel type stick–slip actuator means the motion direction of the stator is parallel to that of the slider. For example, Lee et al. presented a stick–slip actuator

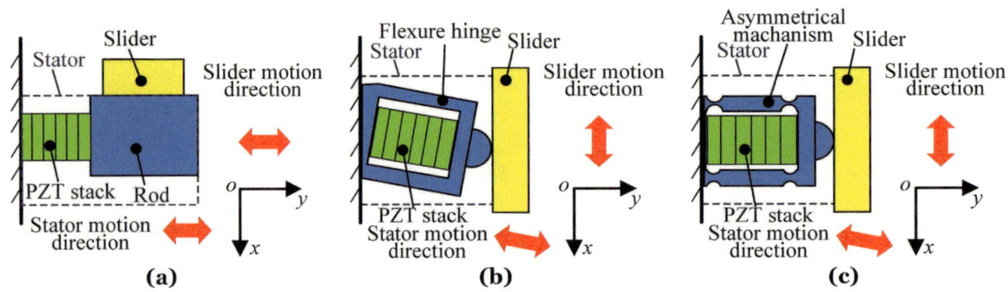

Figure 1.
Schematic diagram of parallel and non-parallel structures. (a) Parallel structure. (b) Non-parallel structure with an angle. (c) Non-parallel structure with asymmetrical flexure hinge.

with the parallel structure using a piezoelectric element and two sliders; and the friction force could be controlled by the spring and the screw [4]. Hunstig et al. proposed a stick–slip actuator with parallel structure, and the friction force was controlled by means of magnets, which could also be predicted by the established friction model [6]. However, the friction force of the parallel type stick–slip actuators cannot be adjusted during the operation, which results in an obvious backward motion. In addition, the velocity and load of such actuators are also seriously restrained.

The piezoelectric stick–slip actuator with non-parallel structure means that there is a movement angle between the stator and the slider. For instance, Li et al. proposed a linear stick–slip actuator with a non-parallel structure based on the lateral motion, which could be generated by a parallelogram-type flexure hinge mechanism and a piezoelectric element [7]. Wen et al. presented a rotary stick–slip actuator with non-parallel structure, and the friction force could be regulated by changing the positive pressure between the slider and stator [9]. Although the velocity and load of the non-parallel type stick–slip actuators are improved relative to the parallel type stick–slip actuators, it is still difficult to meet the actual application requirements. Meanwhile, the backward motions are also generated during the operation. Besides, such actuators are generally structurally asymmetric and perform poor consistency in bidirectional performance.

In this regard, our team focused deeply on the fields of the piezoelectric stick–slip actuators in both structural designs and driving methods. In terms of structural designs, the axial stiffness of the stator is unevenly distributed, and the parasitic motion is generated by introducing the flexure hinge mechanism into the design of the stator, including trapezoid-type piezoelectric stick–slip actuator [10], piezoelectric stick–slip actuators with asymmetrical flexure hinges [11, 12] and mode conversion piezoelectric stick–slip actuator [13]; these structural designs can comprehensively adjust the friction force during the movement of the stick–slip actuator, thereby significantly improving the velocity and load; besides, a coupled asymmetrical flexure hinge mechanism is also developed to achieve the bidirectional motion of the non-parallel type stick–slip actuators [14]. In terms of driving methods, a non-resonant mode smooth driving method (SDM) based on ultrasonic friction reduction is first proposed for the parallel type stick–slip actuator, and the backward motion is restrained during the rapid contraction stage [15]. According to ultrasonic friction reduction, the smaller kinetic friction between the frictional part and the slider can be realized at the resonant frequency. A resonant mode SDM is further developed to improve the performance of the stick–slip actuator [16]. On this basis, the low voltage characteristics and symmetry of the SDM are researched [17, 18]. Finally, a direction-guidance hybrid method (DGHM) is proposed for the non-parallel type stick–slip actuator with flexure hinge mechanism [19], and the superior performance of this excitation method is achieved, especially for high speed.

2. Development of the asymmetric stator with flexure hinge structures

2.1 Trapezoid-type piezoelectric stick-slip actuator

In this work, a trapezoid-type piezoelectric stick–slip actuator adopts a flexure four-bar mechanism and driving foot with trapezoid beam to obtain the oblique motion, which can make the slider move by oblique displacement of flexure hinge mechanism [10], as shown in **Figure 2**.

The structure diagram of the trapezoid-type piezoelectric stick–slip actuator is illustrated in **Figure 2(a)**. The *ox*, *oy*, and *oz* represent the horizontal, axial and vertical directions, respectively. The actuator is mainly comprised of a flexure hinge mechanism, a slider, a piezoelectric stack, a shim block, a preload screw, a base and a preload mechanism. The preload between the flexure hinge mechanism and

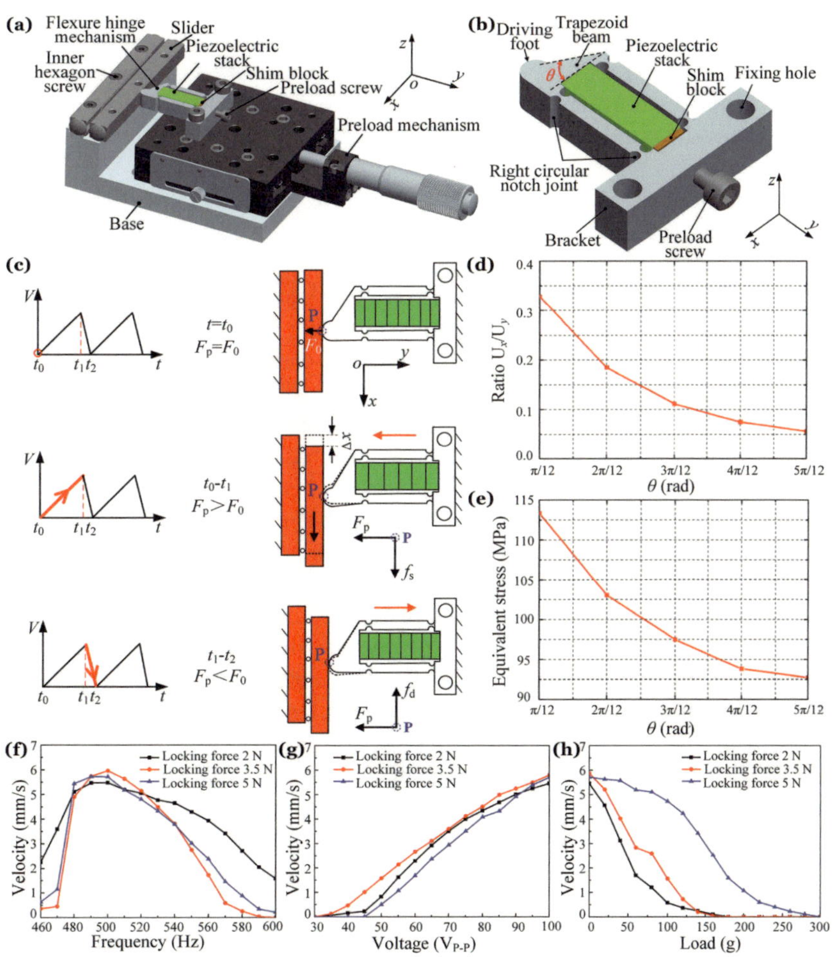

Figure 2.
Research on the trapezoid-type piezoelectric stick–slip actuator. (a) Configuration of the proposed actuator, and (b) the detailed structure of the flexure hinge mechanism. (c) Operation principle of the actuator. (d) and (e) simulation results of the flexure hinge mechanism. (f)–(h) The test results of the actuator. [10].

piezoelectric stack in the initial state is adjusted by the preload screw. The flexure hinge mechanism is preload by the preload mechanism. The needed driving force is generated by the stack which is mounted in the flexure hinge mechanism.

The flexure hinge mechanism is a key component, and its detailed structure is shown in **Figure 2(b)**. The material of the flexure hinge mechanism is AL7075. Four right circular notch joints are employed to the flexure hinge mechanism and thereby the displacement along the axial direction is generated when the piezoelectric stack occurs deformation. It is important that the flexure hinge mechanism utilizes the uneven stiffness distribution of the trapezoid beam to cause needed lateral motion on the driving foot.

The principle of the trapezoid-type stick–slip actuator is depicted in **Figure 2(c)**. In the initial state, when time $t = t_0$, the stack and flexure hinge mechanism keep the original state. The point P is a contact point between the slider and driving foot. The preload force F_0 is equal to F_p generated by the preload mechanism.

1. Slow extension stage ($t_0 \sim t_1$): the stack extends slowly to push the flexure hinge mechanism to expand. In terms of the slider, due to the static friction force f_s between the slider and driving foot, a distance of positive direction is produced on the horizontal axis.

2. Quick contraction stage ($t_1 \sim t_2$): due to the quick contraction of the stack, the flexure hinge mechanism restores to its original condition. And the inertial force makes the condition of the slider remain unchanged.

After the two stages, a small distance of positive direction Δx is generated on the horizontal axis in terms of the slider. The repetition of the above stages makes the actuator generate a long-stroke linear motion.

The lateral displacement of the driving foot can be adjusted by adjusting the angle between the hypotenuse of the trapezoid beam and horizontal axis. To determine the lateral movement on the driving foot, the finite element method (FEM) is used to analyze the angle adjustment process, as shown in **Figure 2(d)** and **(e)**. From the ratio U_x/U_y, a smaller θ rad is needed for better actuation. From the equivalent stress, a larger θ has smaller equivalent stress. Besides, a smaller θ is also essential to meet the requirement of compact size. Finally, $\theta = \pi/6$ rad is selected in this work.

To research the characteristics of the trapezoid-type piezoelectric stick–slip actuator, the prototype and experimental system are established. And a series of experiments are carried out to research the influence of locking force, driving frequency and driving voltage on the performance, as shown in **Figure 2(f)–(h)**.

It is worth noting that the locking force between the driving foot and the slider could change the system stiffness and thereby, the optimal frequency appears the slight change. When the voltage is 100 $V_{p\text{-}p}$, and the frequency is inferior to 500 Hz respectively, the velocity rises with the increase of frequency. When the frequency exceeds 500 Hz, the velocity decreases, which results from that the time of deforming to theoretical length is too short for the stack. It can be seen from **Figure 2(f)** that the maximum velocity is 5.96 mm/s under the locking force of 3.5 N. The velocity increases linearly with an increasing voltage. When the voltage is 100 $V_{p\text{-}p}$, and the locking force is 3.5 N respectively, a maximum velocity of 5.96 mm/s is achieved. Meanwhile, when the voltage is 45.17 $V_{p\text{-}p}$, the minimum starting voltage of the actuator, the displacement resolution of 50 nm is achieved under the locking force of 5 N.

In the initial state, the mass of the slider is 35 g. The standard weight is used to measure the load along the horizontal axis of the actuator. The various loads along the positive direction on the horizontal axis are applied in the slider through a pulley and a steel wire. It is conspicuous that the velocity decreases with the increase in load. The maximum loads of the actuator that can be sustained are 3 N, 1.6 N and 1.6 N when locking forces are 5 N, 3.5 N and 2 N, respectively. The linear relationship between the load and velocity does not emerged, which results from the influence of the assembly errors and the preloading gap.

2.2 Piezoelectric stick-slip actuators using asymmetrical flexure hinges

The trapezoid-type stick–slip actuator developed above can make the slider move by oblique displacement of the flexure hinge mechanism, but it is hard to realize a larger tangential displacement along the motion direction. Therefore, we proposed an idea of using a new four-bar mechanism with different thicknesses of both side flexure hinges, as shown in **Figure 3**. There are considerable improvements in output performance of the actuator based on this design idea.

A four-bar mechanism using asymmetrical flexure hinges is combined with a symmetrical indenter to generate controllable tangential displacement [11]. As shown in **Figure 3(a)**, the stator is mainly comprised of an asymmetrical flexure

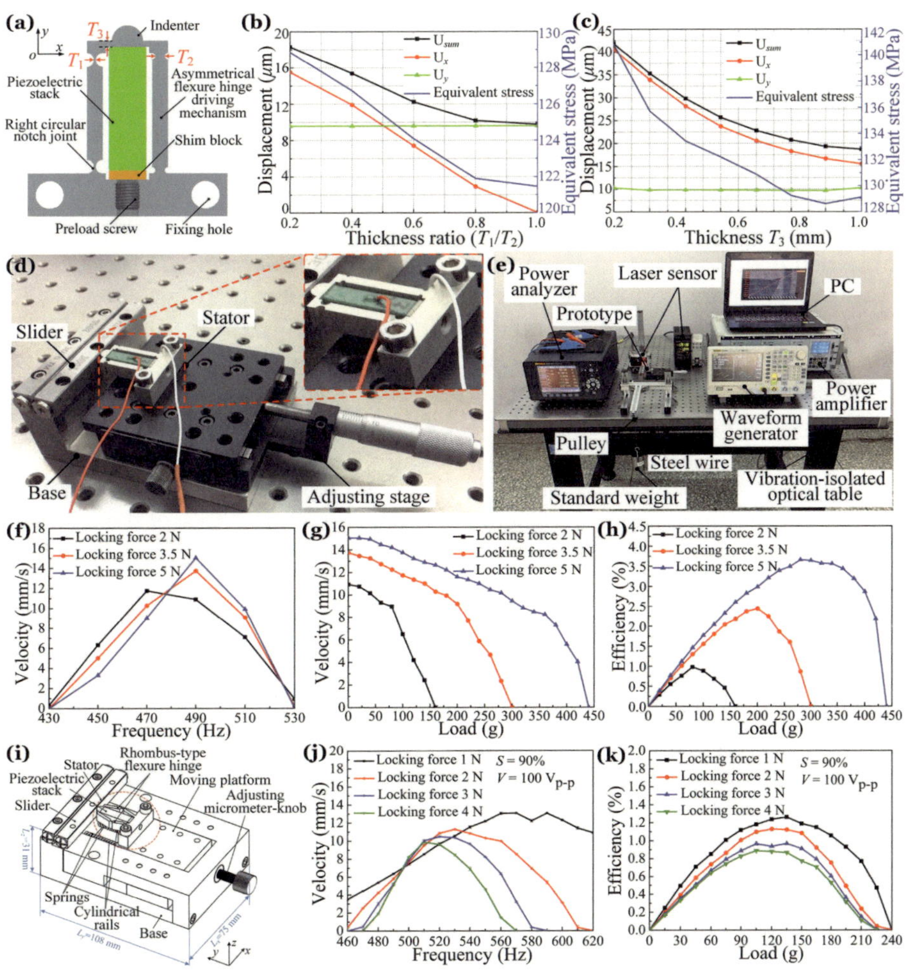

Figure 3.
Research on the piezoelectric stick–slip actuators using asymmetrical flexure hinges. (a) the basic structure of the stator. (b) and (c) simulation results of the asymmetrical flexure hinge mechanism. (d) Prototype and (e) experimental system of the proposed actuator. (f)–(h) The test results of the actuator [11]. (i) The basic structure of the rhombus-type flexure hinge mechanism. (j) and (k) the experimental results of the actuator [12].

hinge driving mechanism, a preload screw, a tungsten chrome steel sheet and a piezoelectric stack. There are right-circle flexure hinges located in both side of the asymmetrical flexure hinge driving mechanism respectively. The thickness of two right-circle flexure hinges on the one side is T_1, and the other side is T_2 ($T_1 \neq T_2$). And the thickness between the stack and the indenter is T_3. By changing the thickness T_1, T_2 and T_3 of the actuators, the oblique movement required by the indenter can be realized.

Deformation of the actuator caused by the different minimum thicknesses of the flexure hinges results in a larger tangential displacement. In the simulation analysis stage, the initial thickness T_1, T_2 and T_3 of the asymmetrical flexure hinge driving mechanism are set as 0.3 mm, 0.3 mm and 4.5 mm, respectively. The simulation results of the asymmetrical flexure hinge driving mechanism under the different ratio of T_1/T_2 and the different thickness of T_3 is shown in **Figure 3(b)** and **(c)**. Based on the results, the mechanism can not only generate the tangential displacement but also change the normal force. Finally, the structure parameters are identified as ratio T_1/T_2 = 0.2 and thickness T_3 = 1 mm.

The actuator mainly consists of a base, a slider, a stator and an adjusting stage, as shown in **Figure 3(d)**. The adjusting stage is used to adjust the preload force between the stator and the slider. The experimental system of the prototype is shown in **Figure 3(e)**. All the experimental equipment is placed on the vibration isolation optical platform.

The experimental results are shown in **Figure 3(f)–(h)**. First, when the voltage is 100 $V_{p\text{-}p}$, the velocity is parabolic with the increase of frequency. When the frequency is 470 Hz, and the locking force is 2 N respectively, the obtained maximum velocity of the actuator is 11.75 mm/s. When the locking forces are 3.5 N and 5 N, the obtained maximum velocities of the actuator under the frequency of 490 Hz are 13.72 and 15.04 mm/s, respectively. Besides, the minimum starting voltage is 20.25 $V_{p\text{-}p}$ at the locking force of 3.5 N, and the displacement resolution is 65 nm. Since frictional force drives the slider to produce linear motion, the improvement of the load capacity may be attributed to the increase of frictional forces as the locking force increased. The maximum loads are 160 g, 300 g and 440 g under locking forces of 2 N, 3.5 N and 5 N, respectively. And the maximum efficiency of the actuator under the locking force of 2 N is 0.98% at the load of 80 g. The maximum efficiency of the actuator under the locking force of 3.5 N is 2.43% at the load of 200 g. The maximum efficiency of the actuator under the locking force of 5 N is 3.66% at the load of 280 g.

Furthermore, a rhombus-type flexure hinge mechanism with an asymmetrical structure is proposed [12], which is easy to produce the parasitic motion of the large stroke. The proposed rhombus-type flexure hinge mechanism can improve the velocity of the actuator with a compact stator. The experimental results show that the prototype achieves a maximum velocity of 13.08 mm/s at a frequency of 570 Hz under the voltage of 100 $V_{p\text{-}p}$; and the maximum efficiency is 1.26% with a load of 135 g.

2.3 Mode conversion piezoelectric stick-slip actuator

A novel stick–slip piezoelectric stick–slip actuator based on mode conversion flexure hinge is proposed which is mainly uses the three chutes of the mode conversion flexure hinge to make the driving foot produce lateral displacement [13], as illustrated in **Figure 4(a)** and **(b)**. Under the influence of the symmetrical waveform, the mode conversion flexure hinge can obtain a lateral displacement and push the slider to move a distance.

The mode conversion flexure hinge is comprised of a bracket, a driving foot, four circular notch joins, a beam and three chutes. The deformation of the

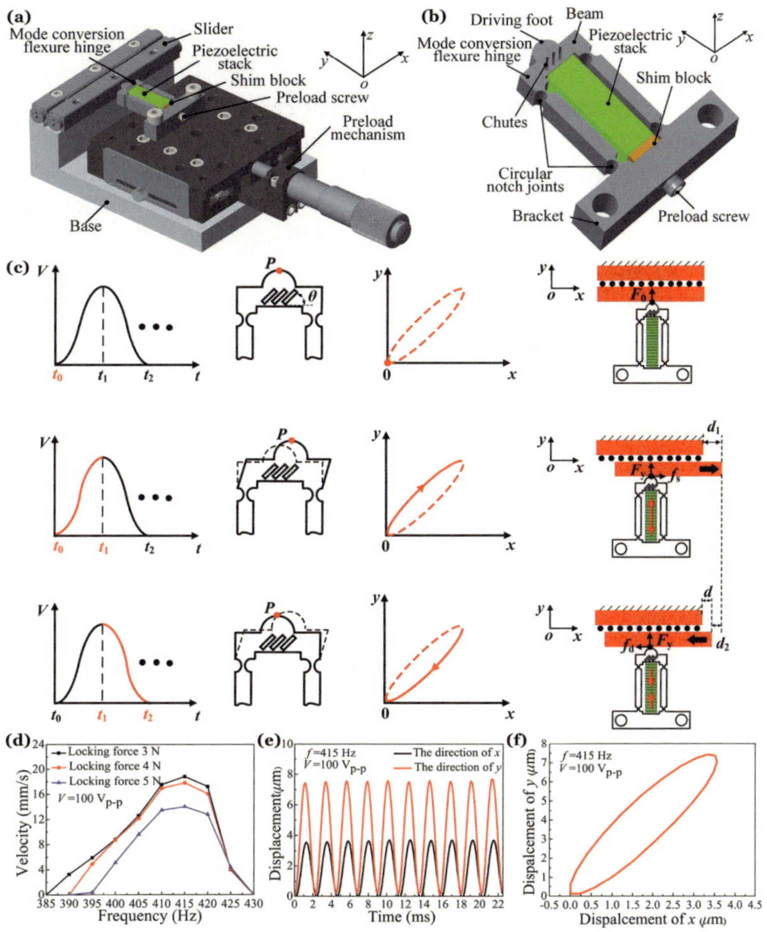

Figure 4.
Research on the mode conversion piezoelectric stick–slip actuator. (a) The detailed structure of the actuator and (b) the stator. (c) Operation principle of the actuator. (d)–(f) The experimental results of the actuator [13].

mode conversion flexure hinge can be produced easily due to four circular notch joins. A symmetrical waveform voltage is chosen to excite the stack. As shown in **Figure 4(c)**, the operation principle of the prototype can be described in the following three steps:

Step 1: no voltage is applied to the stack, and its original length remains. The contact point P is between the driving foot and slider where there is an initial preload force F_0.

Step 2: the point P will generate a displacement of positive directions on both *ox* and *oy* axis with the gradual increase of voltage. The pressure force F_y between the mode conversion flexure hinge and slider improves due to the displacement of the positive direction on the *oy* axis. The slider produces a displacement of the positive direction on the *ox* axis. Within a certain range, as the pressure force F_y increases, the output force of the actuator will be improved. Thereby, the static friction force f_s makes the slider generate a displacement d_1.

Step 3: the stack restores to its original condition with the decrease of voltage. At this moment, kinetic friction force f_d as a resistance force produces between the mode conversion flexure hinge and the slider. The decrease in pressure force F_y makes the force f_d decrease. Then, a slight back displacement d_2 of the slider is produced, which is opposite to the direction of d_1.

Finally, a displacement of the slider d ($d = d_1$-d_2) is produced. By repeating these steps, the prototype can achieve a large motion in the positive direction of the ox axis.

It is an important parameter for the angle θ of chutes to influent its output performance. Different angles of chutes (30°, 45° and 60°) are applied to the proposed mode conversion flexure hinge, which is simulated by the finite element method to obtain an optimal angle parameter. And the displacement U_x of the point P in the positive direction of the ox axis and the maximum equivalent stress with different angles are listed in **Table 1**. The largest deformation of 2.85 μm is obtained for the ox axis displacement U_x of the point P under the angle of 45°, and the maximum equivalent stress is 105.86 MPa.

With the increase of frequency, the velocity of the prototype first increases and then decreases, whose peak can be reached at 415 Hz, as illustrated in **Figure 4(d)**. For the designed actuator, the maximum velocity of 18.84 mm/s, 17.82 mm/s and 14.04 mm/s are obtained respectively under the locking force of 3 N, 4 N and 5 N.

Moreover, the vibrations and the fitting results in xoy plane of the driving foot are shown in **Figure 4(e)** and **(f)**. An ellipse motion track obtained by the driving foot of the actuator is beneficial to the motion of the slider. It can be seen from experiment results that the maximum velocity and efficiency is 18.84 mm/s and 2.07% when the voltage of the symmetrical waveform and the frequency are 100 $V_{p\text{-}p}$ and of 415 Hz, respectively. Besides, the maximum load and the displacement resolution can reach 360 g and 50 nm in terms of the actuator.

2.4 Bidirectional piezoelectric stick-slip actuator

The asymmetric flexure hinges are generally structurally asymmetric, making it difficult to achieve good consistency in bidirectional output performance. In this work, a coupled asymmetrical flexure hinge mechanism is developed to realize the bidirectional motion of the piezoelectric stick–slip actuator [14], as shown in **Figure 5**.

The bidirectional stick–slip actuator uses coupled asymmetrical flexure hinge mechanisms and symmetrical indenter to generate the controllable tangential displacement. As shown in **Figure 5(a)**, the static simulation of the FEM is used to analyze the deformation of flexure hinges with different sizes. Since a larger oblique displacement will provide better output performance, A_1 = 1 mm and $K = A_2/A_3 = 0.2$ with the larger oblique displacement are selected for the final Structure design.

In this section, an MDR-based model is proposed to display the contact and frictional force between the slider and indenter for stick–slip actuators.

The framework of the MDR consists of two preliminary steps to be performed. As shown in **Figure 5(c)**, a one-dimensional profile $g(x)$ and the elastic half-space by an elastic foundation which are in possession of normal stiffness k_y and tangential stiffness k_x is adopted to replace the three-dimensional profile $y = f(x, z)$.

Angle of the chutes	Deformation (μm)	Equivalent stress (MPa)
30°	2.77	106.63
45°	2.85	105.86
60°	2.42	103.32

Table 1.
The simulation results of the proposed actuator.

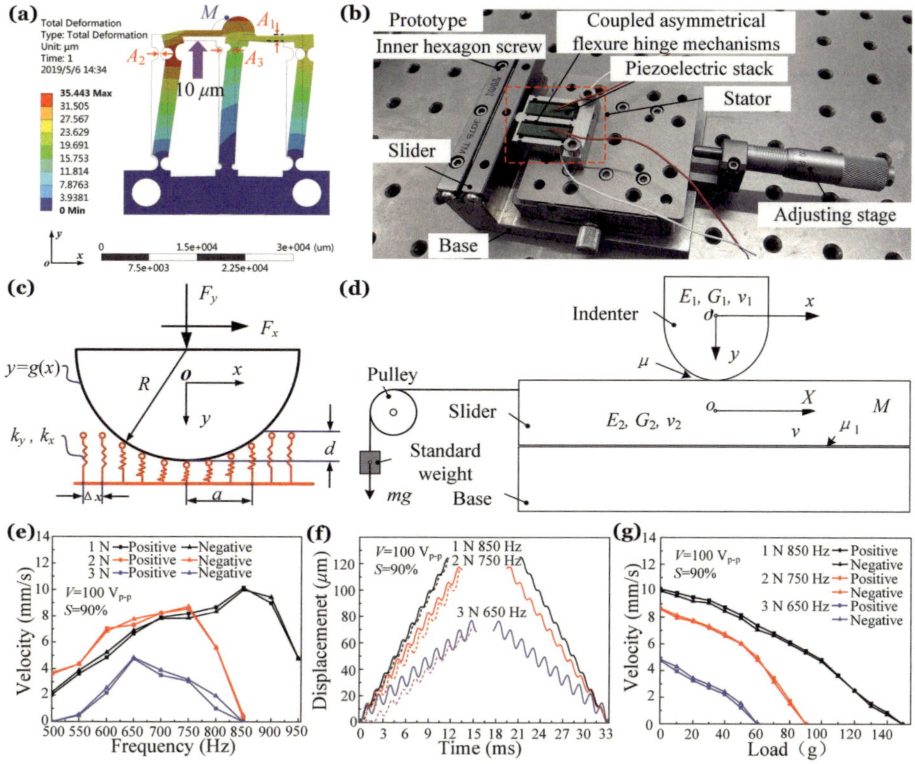

Figure 5.
Research on the bidirectional piezoelectric stick–slip actuator. (a) The static simulation deformation of the flexure hinges. (b) The prototype of the actuator. (c)–(d) MDR model. (e)–(g) The experimental results of the actuator [14].

The stiffness of every individual spring can be expressed as follows:

$$k_y = E^* \Delta x, k_x = G^* \Delta x \tag{1}$$

Δx represents the distance between the springs of elastic foundation, and the effective elastic modulus is defined as:

$$\frac{1}{E^*} = \frac{1-\nu_1^2}{E_1} + \frac{1-\nu_2^2}{E_2} \tag{2}$$

$$\frac{1}{G^*} = \frac{1-\nu_1^2}{G_1} + \frac{1-\nu_2^2}{G_2} \tag{3}$$

$$G_{1,2} = \frac{E_{1,2}}{2\left(1+\nu_{1,2}\right)} \tag{4}$$

E_1 G_1 and ν_1 represent the Young's modulus of the indenter, its shear modulus and the Poisson's ratio, respectively. E_2, G_2 and ν_2 are the corresponding material parameters of the half-space.

Figure 5(d) shows the schematic diagram of the experimental system. Here, the friction coefficient of the slider in contact with the base is set to μ_1. The friction coefficient that cylindrical indenter drives the slider is set as μ. The indenter is made of a material with parameters of E_1, G_1 and ν_1, and elastic parameters of the slider material are E_2, G_2 and ν_2.

Further modeling is based on experimental data. Newton's second law then yields as follows:

$$M\ddot{X} = F_x - F_{base} - mg \tag{5}$$

F_x represents the frictional force between the indenter and slider. With the movement of the slider, a sliding friction force F_{base} goes up between the slider and the base. It is calculated as

$$F_{base} = \mathrm{sgn}\left(\dot{X}\right)\mu_1 F_y \tag{6}$$

Since in the numerical calculation, we can only use a discrete model as follows. In the circumstances, the value of the elongation of spring adds relative displacement $\Delta\tilde{x}$ that can be calculated as follows:

$$\Delta\tilde{x} = \left[x(t_i + \Delta t) - x(t_i)\right] - \left[X(t_i + \Delta t) - X(t_i)\right] \tag{7}$$

where t_i and Δt are the incrementally increasing time and the step of numerical time integration of Eq. (5) (time increment value).

The function $g(x)$ is found for a cylinder of finite length that is indented into a half-space. The form of this function is defined as follows:

$$g(x) = \beta\frac{L^2}{R}\left(\frac{\alpha L}{x} + 1\right)\exp\left(-\frac{\alpha L}{x}\right) \tag{8}$$

where L and R are the cylinder length and its radius, respectively. In addition, the ratio L/R determines the values of the parameters α and β.

The velocity is measured at a voltage with the saw-tooth wave of 100 $V_{p\text{-}p}$, and the duty ratio "*S*" is 90%, as shown in **Figure 5(e)**. As the frequency changes, the velocity trend shows a parabolic. And the different locking forces have different optimal driving frequencies. The *x* direction is taken as the moving direction of the slider. When the frequency is 850 Hz and locking force is 1 N, the maximum velocities in the positive direction and negative direction are 10.14 mm/s and 9.99 mm/s. When the frequency is 750 Hz and locking force is 2 N mm/s, the maximum velocities of the actuator are 8.56 and 8.72 mm/s in the positive direction and negative direction. When the frequency is 650 Hz and locking force is 3 N, the maximum velocities of the actuator are about 4.75 and 4.84 mm/s in the positive direction and negative direction.

As shown in **Figure 5(f)**, the solid lines and the dotted lines are the displacement curves and the simulation displacement curve of the slider under different locking forces. When the voltage is in a stage of rapid decrease, the dynamic friction will be generated between the indenter and the slider. Meanwhile, the difference in the direction of the dynamic friction and the slider results in the production of the backward motion. Therefore, every step can produce a backward motion. Although it can be seen from the dotted line that the slider moves in forwarding and backward directions, the average velocity that is larger than zero will result in a net directional motion.

Experimental results indicate that the maximum loads tested in the experiment are 0.6 N, 0.9 N and 1.5 N under locking forces of 1 N, 2 N and 3 N, respectively. Furthermore, when the locking force and the minimum starting voltage are 2 N and 31.9 $V_{p\text{-}p}$ respectively, the forward and reversed displacement resolutions of the actuator are 91.1 nm and 74.4 nm. When the voltage and the frequency are 100 $V_{p\text{-}p}$ and 850 Hz, the maximum output velocity and the maximum load of the actuator in the positive x-direction is 10.14 mm/s and 1.5 N. When the load is 90 g, a locking force is 5 N, and the velocity is 5.48 mm/s, the actuator can reach the maximum efficiency of 0.57%.

3. Development of the hybrid excitation methods

3.1 Smooth driving method using ultrasonic friction reduction

3.1.1 Operation principle of smooth driving method

Figure 6(a) shows the working principle of the piezoelectric stick–slip actuator by traditional driving method (TDM), which includes the slow extension and rapid contraction stages of the traditional saw-tooth driving wave. As illustrated in **Figure 6(b)**, the smooth driving method (SDM) is realized by a composite wave, in which the composite wave includes a saw-tooth driving wave (SD-wave) and a sinusoidal friction regulation wave (SFR-wave), in which the SFR-wave can adjust the friction force between the frictional part and slider in resonant mode (RSFR-wave) or non-resonant mode (NSFR-wave) [15]. From the time A to B, the composite wave is equivalent to one part of the SD-wave. The piezoelectric element is excited to extend slowly, and the slider will move a distance together with the frictional part in the axial direction by a static friction force f_s. From the time B to C, the composite wave is composed by an SFR-wave and another part of the SD-wave, which means that the ultrasonic friction reduction is introduced into the composite wave to decrease the kinetic friction force f_d. Thus, the backward motion of the

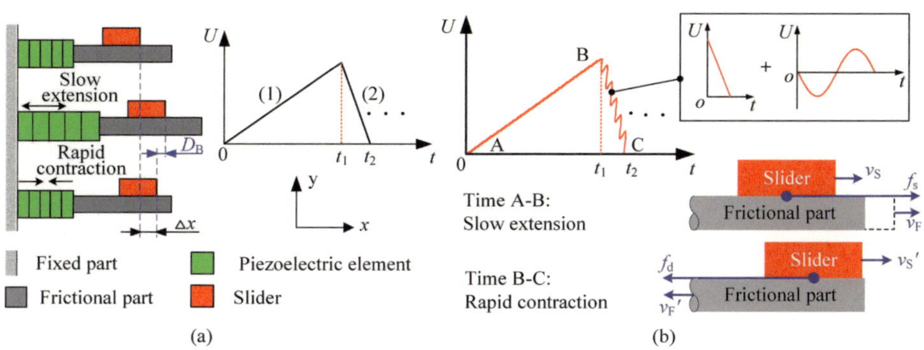

Figure 6.
Operation principle. (a) Principle of the traditional driving method. (b) Principle of the smooth driving method [15].

stick–slip actuator can be restrained. The continuous output movement is gained by repeating the above motion process. The reverse motion can be realized by reflecting the symmetry of the SDM.

3.1.2 Research on non-resonant mode smooth driving method

In this work, a non-resonant mode SDM is proposed to restrain the backward motion [15], which is realized by applying NSFR-wave to rapid deformation stage of the SD-wave. The specific work is as follows: the prototype includes a piezoelectric element, a frictional part, a slider and a preload mechanism, see **Figure 7(a)**. **Figure 7(b)** shows the experimental system of the prototype.

Figure 7(c) shows the amplitudes of the frictional part under the TDM and SDM. The frequency of the SD-wave is set as 800 Hz, and the voltage and frequency of the NSFR-wave are set as 4 $V_{p\text{-}p}$ and 40 kHz. It can be seen that the amplitude of the frictional part is not affected obviously under the excitation of the SDM. **Figure 7(d)** illustrates the displacement curves of the prototype with a load of 4 g. Compared with the TDM, the better performance is achieved by the SDM. To quantify the restraint degree of backward motion, the backward rate (denoted by ζ) is defined by a ratio of backward distance D_B and driving distance D_S. The results indicate that the backward rate is decreased markedly by 83%. The relationship between the velocity and NSFR-wave voltage is plotted in **Figure 7(e)**. The velocity with the NSFR-wave voltage of 0 $V_{p\text{-}p}$ represents the velocity under the excitation of the TDM; it is found that the velocity increases linearly with an increasing NSFR-wave voltage, and the velocity is obviously improved using the SDM relative to the TDM. **Figure 7(f)** shows the relationship between the velocity and voltage. The results indicate that there is a linear increasing tendency between the velocity and voltage.

3.1.3 Research on resonant mode smooth driving method

On the basis of our previous work, a resonant mode SDM is further developed to improve the performances of the actuators [16]. The resonant mode SDM is realized by applying a RSFR-wave to rapid deformation stage of the SD-wave. The relative

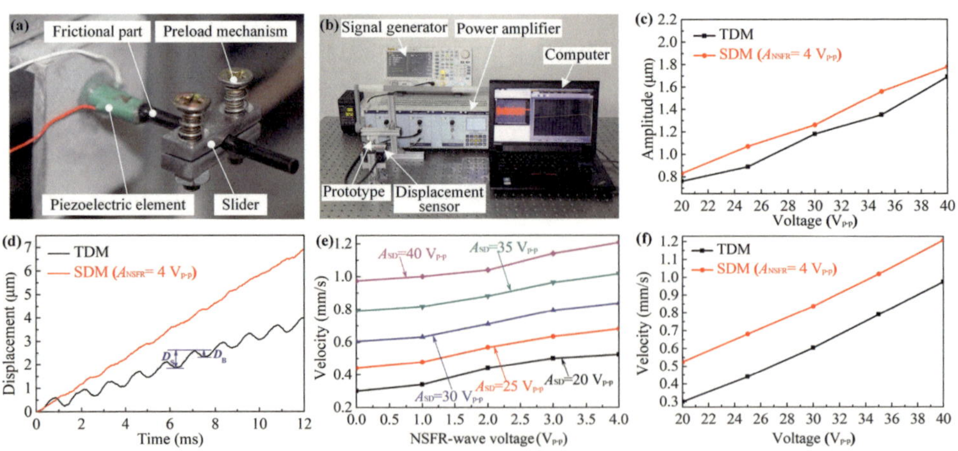

Figure 7.
Non-resonant mode SDM. (a) Prototype of the stick–slip actuator. (b) Experimental system. (c) Amplitude of the friction part excited. (d) Displacement curves of the prototype. (e) Velocity versus the NSFR-wave voltage. (f) Velocity versus the voltage [15].

research is as follows: the modal analysis of the stator is done by the FEM, aiming at obtaining the 1st longitudinal vibration mode, as illustrated in **Figure 8(a)**; and the theoretical result indicates that the resonant frequency of 38.55 kHz is obtained, which is considered the exciting frequency of the RSFR-wave.

Figure 8(b) shows the velocity versus the frequency of the RSFR-wave; the driving voltage of the SD-wave is set as 30 $V_{p\text{-}p}$, and the RSFR-wave voltage is selected as 6 $V_{p\text{-}p}$; the velocity of the actuator increases first and then decreases with an increasing frequency, and the maximum velocity is 0.41 mm/s under 39 kHz. Therefore, the RSFR-wave frequency of 39 kHz is chosen in subsequent experiments; there is a slight deviation between the test result and simulation result; the reasons may be caused by material errors between the model and the prototype, the omission of adhesive layers and the machining errors; besides, the comparative experiments of the SDM are also developed under the condition of 36 kHz and 42 kHz.

Figure 8(c) shows the relationship between the displacement and time; the displacement using the TDM shows a fluctuated curve and the backward motion is seen in every step; conversely, the backward motion is restrained by the SDM; the displacement increases linearly with an increasing time; the linearity is better when the frequency of RSFR-wave is 39 kHz, because the smaller kinetic friction is realized relative to the 36 kHz and 42 kHz. **Figure 8(d)** shows the relationship between the velocity and voltage; the velocity approximately follows a linear increased tendency with the voltage; the velocity using the TDM is 0.163 mm/s under 30 $V_{p\text{-}p}$; meanwhile, the velocities through the SDM are 0.332 mm/s, 0.403 mm/s and 0.370 mm/s (i.e. 36 kHz, 39 kHz, and 42 kHz), and the corresponding velocities can be effectively improved by 103.68%, 147.23% and 126.99% compared with the TDM.

Figure 8(e) expresses the variation of the velocity with an increasing load; the standard weight is used to investigate the load characteristics, which are added to the slider. The measure results indicate that the velocity decreases with an increasing load, and the load can reach 125 g utilizing the TDM; meanwhile, the maximum load excited by the SDM are 290 g, 360 g, and 315 g under 36 kHz, 39 kHz and 42 kHz; besides, under the same load, a higher velocity by the SDM is achieved at the RSFR-wave frequency of 39 kHz. **Figure 8(f)** shows that the relationship between the driving capacity and load; the driving capacity of the actuator increases first and then decreases with an increasing load, and the maximum

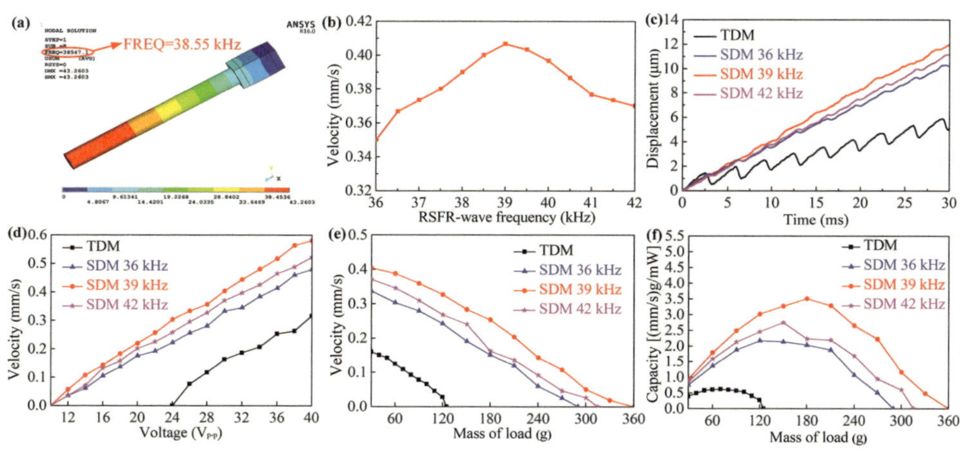

Figure 8.
Resonant mode SDM. (a) Modal analysis result of the stator. (b) Velocity versus the RSFR-wave frequency. (c) Displacement versus the time. (d) Velocity versus the voltage. (e) Velocity versus the mass of the load. (f) Driving capacity versus the mass of the load [16].

driving capacity using the TDM is 0.62 [(mm/s)g/mW] at 70 g. In contrast, the maximum driving capacities using the TDM (i.e. 36 kHz, 39 kHz and 42 kHz) are 2.18 [(mm/s)g/mW], 3.51 [(mm/s)g/mW] and 2.75 [(mm/s)g/mW] under 120 g, 180 g and 150 g; the driving capacities based on the SDM can be increased by 251.61%, 466.13% and 343.55% under the SFR-wave frequencies of 36 kHz, 39 kHz and 42 kHz.

3.1.4 Research on low voltage characteristics of smooth driving method

In this work, the low voltage characteristics of the SDM is researched [17]. **Figure 9(a)** and **(b)** show the relationships between the velocity and voltage in forward and reverse directions; the experiments are carried out under the same load of 4 g, and the frequencies of the SD-wave and the SFR-wave are 800 Hz and 40 kHz; the voltages of the SDM and TDM are increased from 2 $V_{p\text{-}p}$ to 10 $V_{p\text{-}p}$; the forward and reverse motions are changed by modifying the order of slow and rapid movements as 90% or 10%; the results indicate that the velocity of the actuator increase obviously as the voltage goes up. It also can be seen that the minimum input voltage is reduced from 3 $V_{p\text{-}p}$ to 2 $V_{p\text{-}p}$ under the SDM; there are the similar variation relations in the forward and reverse motions.

Figure 9(c) and **(d)** show the performances of the prototype with a load of 4 g by the SDM, in which the SDM is excited by the SD-wave voltage of 8 $V_{p\text{-}p}$ and SFR-wave of 1.6 $V_{p\text{-}p}$, and the voltage of the TDM is also 8 $V_{p\text{-}p}$. The results indicate that the prototype can achieve a stable operation at low voltage, and the backward motion is observed in every step; and the forward average effective displacement D_e is improved by 64%; the reverse average effective displacement D_e is improved by 55%; thus, the output displacement characteristics of the prototype is improved at a lower input voltage.

Figure 9(e) and **(f)** show the load characteristics of the actuator excited by the SDM and TDM, respectively; the output velocities of the actuator decrease as the load go up; compared with the TDM, the velocities can be improved obviously under the same load condition; the maximum load capacity of the actuator is 24 g when the voltages are 8 $V_{p\text{-}p}$ and 1.6 $V_{p\text{-}p}$, respectively, which is 2.4 times the load capacity of the actuator under the TDM. Therefore, the designed actuator excited by the SDM can achieve a larger load capacity.

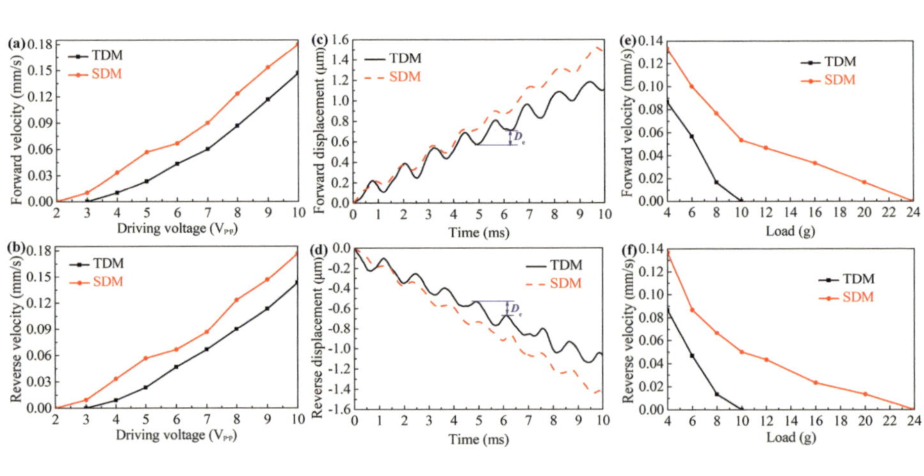

Figure 9.
Low voltage characteristics of the SDM. (a) Forward velocity versus the voltage. (b) Reverse velocity versus the voltage. (c) Forward displacement. (d) Reverse displacement. (e) Forward load characteristics. (f) Reverse load characteristics [17].

3.1.5 Research on symmetry of smooth driving method

In this work, the symmetry of the SDM is also studied in detail [18]. **Figure 10(a)** and **(b)** show the relationship between the symmetry and velocity under the TDM and the SDM with an initial slider mass of 30 g; the voltage and frequency are 10 $V_{p\text{-}p}$ and 800 Hz. In terms of the TDM, the actuator under the symmetry of 50–80% and 20–50% cannot work properly, because the effective displacement cannot be generated; the maximum velocities are 0.21 mm/s and 0.20 mm/s under 95% and 5%; the symmetries between 80–95% and 5–20% are considered as the ideal working ranges. In terms of the SDM, the velocities under 95% and 5% can be improved obviously relative to the TDM, and the velocities can reach 0.24 mm/s and 0.23 mm/s. The maximum velocities in the forward and reverse directions can be achieved at 50%, and the corresponding velocities are 0.41 mm/s and 0.39 mm/s. It is found that the effective symmetry of the SDM is obviously widened relative to the TDM.

Figure 10(c) and **(d)** show the relationship between the load and velocity under the different symmetries, such as 90%, 70%, 50%, 30% and 10% in forward and reverse directions; the velocity of the actuator decreases with the increase of load; the velocities under 50%, 70% and 90% are almost equal at 110 g; when the load is less than 110 g, the maximum velocity can be achieved under 50%; when the load is greater than 110 g, the maximum velocity can be realized under 90%.

It is found that the asymmetrical SDM under 90% achieves the better load capacity when the SD-wave voltage is 10 $V_{p\text{-}p}$ for 800 Hz and the SFR-wave is 2 $V_{p\text{-}p}$ for 39 kHz; the actuator in the reverse direction can also achieve the similar load characteristics; the larger stiffness is obtained under 90% and 10%.

The relationship between the load and driving capacity is illustrated in **Figure 10(e)** and **(f)**; the driving capacities of the actuator can reach 7.92 [(mm/s)g/mW] and 6.77 [(mm/s)g/mW] under 70% and 90%; meanwhile, the driving capacities of the actuator can reach 8.42 [(mm/s)g/mW] and 7.11 [(mm/s)g/mW] under 30% and 10%. The results indicate that the symmetrical SDM such as 50% can achieve larger driving capacity relative to the asymmetrical SDM, such as 90%, 70%, 30% and 10%.

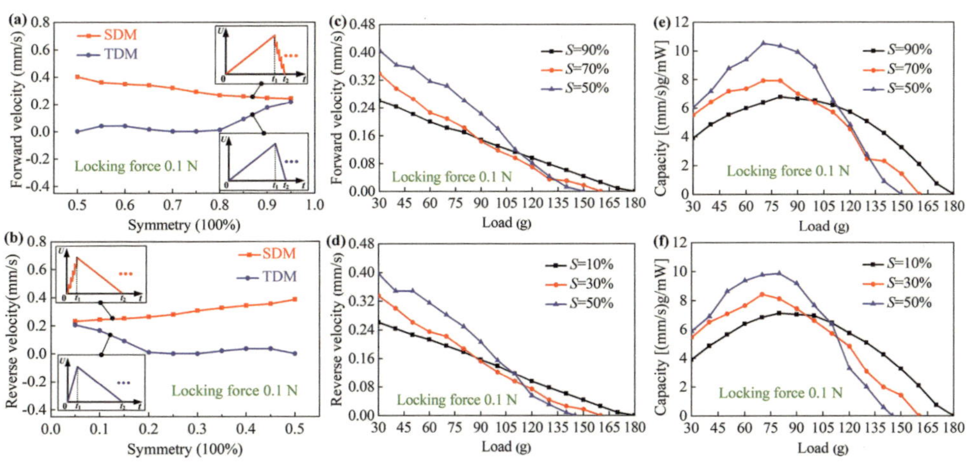

Figure 10.
Symmetry of the SDM. (a) Forward velocity versus the symmetry. (b) Reverse velocity versus the symmetry. (c) Forward velocity versus to the load under 50%, 70% and 90%. (d) Reverse velocity versus to the load under 10%, 30% and 50%. (e) Forward driving capacity versus the load under 50%, 70% and 90%. (f) Reverse driving capacity versus the load under 10%, 30% and 50% [18].

3.2 Direction-guidance hybrid method for high speed

3.2.1 Operation principle of direction-guidance hybrid method

To improve the speed of the stick–slip actuators, a direction-guidance hybrid method (DGHM) is proposed in this work [19]. The DGHM is a synthetic composite waveform, as shown in **Figure 11(a)**, which is mainly composed of a direction-guidance (DG) waveform and a resonance drive (RD) waveform. The stator excited by the DG waveform to produce a pre-deformation with a lateral motion, which is of great significance to the output characteristics of the actuator. The direction of the slider can be guided by the lateral motion of the stator; besides, the locking force can be adjusted to achieve better output performance. After that, the high speed performance of the actuator can be achieved at a resonant frequency of the RD waveform.

The operation principle of the actuator under the DGHM is shown in **Figure 11(b)**. The specific operations are as follows:

Stage I: at time t_0, the piezoelectric stack without input voltage, the length of the stack is l_0, and the slider and the stator are at the initial position. Due to the existence of the initial locking force, there is a pair of equal and opposite interaction forces (F_{I0} and F_{R0}) between the driving foot and the slider.

Stage II: the length of the stack extends to l_1 quickly excited by the DG waveform at time t_1 so that stator is produced an oblique deformation. The position of the driving foot is moved from position ① to position ②, which causes a pre-deformation of the stator. At the same time, the slider will produce a displacement of Δd_1 along the *ox* axis with the help of the driving foot. In addition, the interaction force becomes F_{I1} and F_{R1} after incrementing ΔF ($|F_{I1}| = |F_{R1}| = |F_{I0}| + |\Delta F|$), which can further increase the load capacity and horizontal thrust of the actuator within a proper range of regulation.

Stage III: the stack is excited by the DGHM at time t_2. The excitation voltage of the DG waveform is kept constant. The RD waveform is composed of high frequency sine waveforms with the period of T_R. The stack undergoes an elongation and contraction between the lengths of l_1 and l_2 in each cycle. Correspondingly, the driving foot vibrates rapidly between the positions of ② and ③. During the time of T_R, a displacement Δd_2 of the slider is produced along the *ox* axis. The velocity of

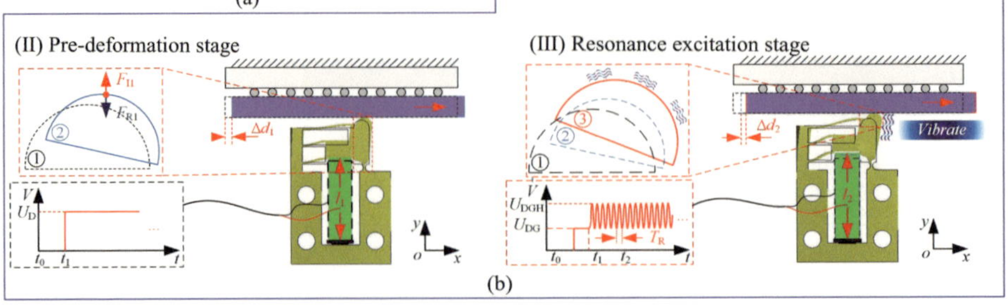

Figure 11.
Operation principle of the DGHM. (a) Schematic of the DGHM. (b) Specific operations of the DGHM [19].

the actuator is v ($v = f_R \times \Delta d_2$), where f_R is the frequency of the RD waveform. Even at a low voltage, the actuator can achieve a high speed by the DGHM.

3.2.2 Research on direction-guidance hybrid method

The related work based on the DGHM is as follows [19]. As shown in **Figure 12(a)**, an actuator is fabricated to verify the feasibility of the DGHM. In the first place, the influence of the DG waveform with different amplitudes of the driving foot is analyzed by the FEM. The point P is chosen as the reference point, which represents the contact point between the slider and the driving foot, as shown in **Figure 12(c)**. **Figure 12(b)** simulates the equivalent stress of the stator and deformation of the point P when the piezoelectric stack inputs different displacements driven by the DG waveform. The displacement of the point P increases linearly in the ox and oy axes directions with the increase of the input displacement, which verifies the effectiveness of the DG waveform. Besides, the equivalent stress of the stator is 202.46 MPa with the input of 10 μm. Afterwards, the modes of the stator are analyzed to make the actuator operate at a resonant frequency. As shown in **Figure 12(c)**, when the frequency of RD waveform is at fourth mode (around 30.57 kHz), the proposed actuator has the potential to achieve higher speed.

The actuator is manufactured and the test system is built, as shown in **Figure 12(d)**. Firstly, the effect of the voltage of DG waveform on the displacement of the driving foot is explored. As shown in **Figure 12(e)**, the driving foot can achieve displacement along

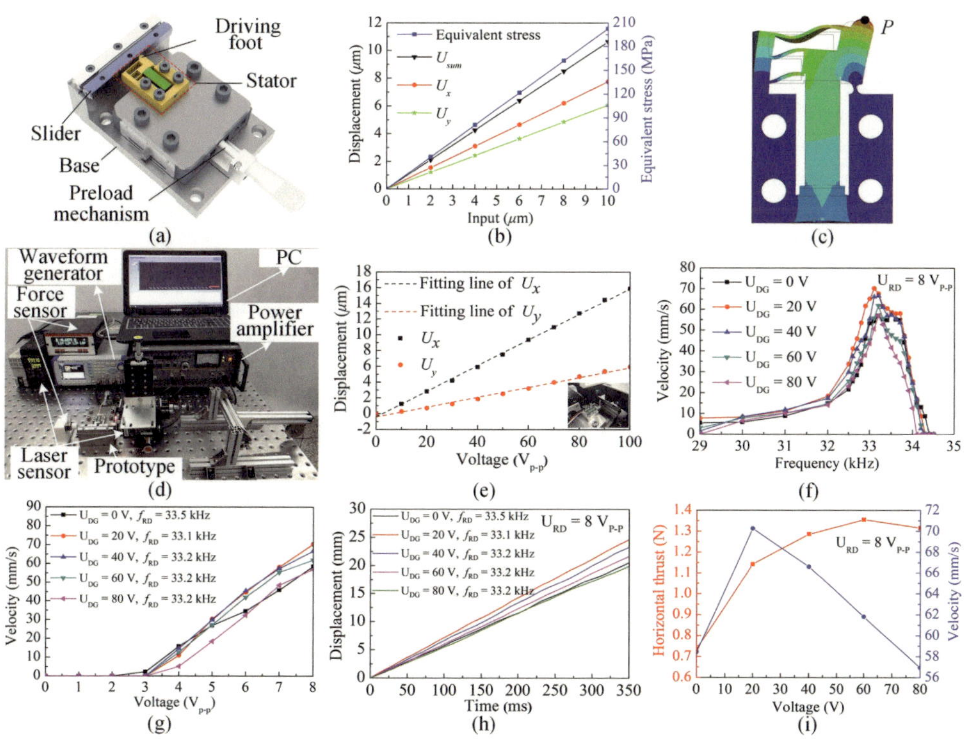

Figure 12.
Research on DGHM. (a) Configuration of the proposed design. (b) Simulation of the stator by FEM. (c) Fourth mode of the stator. (d) The experimental system and the prototype. (e) The relationship between the displacement of the reflector and the voltage of DG waveform. (f) The relationship between driving frequency of DGHM and output velocity. (g) The relationship between the velocity and the voltage of RD waveform under the different voltages of DG waveform. (h) The displacement of the prototype at different voltages of DG waveform. (i) The horizontal thrust and the velocity versus at the voltages of DG waveform under the DGHM [19].

the positive direction of the *ox* and *oy* axes. With the increase of the voltage of the DG waveform, the displacements in both directions have a good linear relationship, which shows that the actuator can produce a pre-deformation by the aid of the proposed excitation method.

Under the condition of an initial preload force of 0.5 N, a series of experiments are carried out to explore the output characteristics of the proposed actuator. As shown in **Figure 12(f)**, the output speed and the frequency of the RD waveform are studied under different voltage of the DG waveform to determine the optimal driving frequency of the RD waveform. At different voltages of the DG waveform, the optimal frequencies of the RD waveform are around 33 kHz, which is close to the frequency simulated by FEM. When the DG waveform is 20 V and the RD waveform is 8 $V_{P\text{-}P}$ and 33.1 kHz, the actuator reaches the maximum velocity of 70.26 mm/s.

Under different DG waveform voltages, the velocities of the actuator versus voltages of the RD waveform are measured. It can be seen from **Figure 12(g)** that the velocity increases as the RD waveform voltage. The actuator starts to move as the voltage of RD waveform is about 3 $V_{p\text{-}p}$. At the voltage of 8 $V_{p\text{-}p}$, the velocity reaches the maximum and it can be further improved by increasing the voltage of the RD waveform. As shown in **Figure 12(h)**, the displacement curve at different voltages of DG waveform shows that the DGHM can make the actuator move stably. As shown in **Figure 12(i)**, the horizontal thrust and velocity at a different voltage of the DG waveform are explored which shows that the actuator can achieve the large horizontal thrust or the relatively high velocity under the DGHM. Although the initial locking force is 0.5 N, the actuator can achieve the maximum horizontal thrust of 1.36 N when the voltage of DG waveform is 60 V.

4. Conclusions

A comprehensive work of piezoelectric stick–slip actuators was presented to improve their output characteristics. This work divided our team's stick–slip actuators into two categories based on the research ideas, including structural designs and driving methods.

In terms of structural designs, the trapezoid-type actuator increased static friction force in slow extension stage and decreased kinetic friction force in rapid contraction stage by lateral motion, which improved the forward performance. The actuators with asymmetrical flexure hinges widened the structure forms of the actuator and improved the velocity, load and efficiency. The mode conversion actuator could be used in precision positioning platform and ultra-precision machining due to its nanometer resolution. Besides, an actuator with a coupled asymmetrical flexure hinge mechanism realized the bidirectional motion.

In terms of driving methods, a non-resonant mode smooth driving method (SDM) based on ultrasonic friction reduction was proposed, and the backward motion was restrained during the rapid contraction stage. Compared with non-resonant mode SDM, the resonant mode SDM achieved a higher velocity, larger load and smoother displacement. Besides, the proposed SDM achieved better low voltage characteristics and widened the symmetry relative to the traditional driving method. Finally, a direction-guidance hybrid method (DGHM) could achieve superior performance, especially for high speed.

Acknowledgements

This work was financially supported by the Jilin Province Science and Technology Development Plan Item (No. 20150312006ZG), the Postdoctoral

Science Foundation of China (No. 2015 M571356), the Key Projects of Science and Technology Development Plan of Jilin Province (No. 20160204054GX), the Project of Industrial Technology Research and Development of Jilin Province Development and Reform Commission (No. 2019C037-6) and the Science and Technology Development Plan of Jilin Province (Nos. 20190201108JC and 20200201057JC).

Conflict of interest

The author(s) declared no potential conflicts of interest with respect to the research, authorship, and/or publication of this article.

Author details

Tinghai Cheng[1,2*], Xiaosong Zhang[1], Xiaohui Lu[1], Hengyu Li[3], Qi Gao[2] and Guangda Qiao[1]

1 School of Mechatronic Engineering, Changchun University of Technology, Changchun, China

2 Beijing Institute of Nanoenergy and Nanosystems, Chinese Academy of Sciences, Beijing, China

3 State Key Laboratory of Robotics and System, Harbin Institute of Technology, Harbin, China

*Address all correspondence to: chengtinghai@binn.cas.cn

IntechOpen

© 2021 The Author(s). Licensee IntechOpen. This chapter is distributed under the terms of the Creative Commons Attribution License (http://creativecommons.org/licenses/by/3.0), which permits unrestricted use, distribution, and reproduction in any medium, provided the original work is properly cited.

References

[1] Hunstig M. Piezoelectric inertia motors—a critical review of history, concepts, design, applications, and perspectives. Actuators. 2017;6:35. DOI: 10.3390/act6010007

[2] Qin F, Tian L, Huang H, Wang J, Liang T, Zu X, Zhao H. Actively controlling the contact force of a stick-slip piezoelectric linear actuator by a composite flexure hinge. Sensors and Actuators a-Physical. 2019;299:10. DOI: 10.1016/j.sna.2019.111606

[3] Guo Z, Tian Y, Zhang D, Wang T, Wu M. A novel stick-slip based linear actuator using bi-directional motion of micropositioner. Mechanical Systems and Signal Processing. 2019;128:37-49. DOI: 10.1016/j.ymssp.2019.03.025

[4] Lee J, Kwon W, Kim K, Kim S. A novel smooth impact drive mechanism actuation method with dual-slider for a compact zoom lens system. Review of Scientific Instruments. 2011;82:8. DOI: 10.1063/1.3624701

[5] Peng J, Chen X. Modeling of piezoelectric-drive stick-slip actuator. IEEE/ASME Transactions on Mechatronics. 2011;16:394-399. DOI: 10.1109/tmech.2010.2043849

[6] Hunstig M, Hemsel T, Sextro W. Modelling the friction contact in an inertia motor. Journal of Intelligent Material Systems and Structures. 2013;24:1380-1391. DOI: 10.1177/1045389X12474354

[7] Li J, Zhou X, Zhao H, Shao M, Hou P and Xu X. Design and experimental performances of a piezoelectric linear actuator by means of lateral motion. Smart Materials and Structures. 2015;24:065007. DOI: 10.1088/0964-1726/24/6/065007

[8] Wang S, Rong W, Wang L, Pei Z, Sun L. Design, analysis and experimental performance of a bionic piezoelectric rotary actuator. Journal of Bionic Engineering. 2017;2:348-355. DOI: 10.1016/s1672-6529(16)60403-1

[9] Wen J, Ma J, Zeng P, Cheng G & Zhang Z. A new inertial piezoelectric rotary actuator based on changing the normal pressure. Microsystem Technologies. 2013;19:277-283. DOI: 10.1007/s00542-012-1684-9

[10] Cheng T, He M, Li H, Lu X, Zhao H, Gao H. A novel trapezoid-type stick–slip piezoelectric linear actuator using right circular flexure hinge mechanism. IEEE Transactions on Industrial Electronics. 2017;64:5545-5552. DOI: 10.1109/TIE.2017.2677318

[11] Gao Q, He M, Lu X, Zhang C, Cheng T. Simple and high-performance stick-slip piezoelectric actuator based on an asymmetrical flexure hinge driving mechanism. Journal of Intelligent Material Systems and Structures. 2019;30:2125-2134. DOI: 10.1177/1045389X19862376

[12] Gao Q, Li Y, Lu X, Zhang C, Zhang X, Cheng T. A piezoelectric stick-slip linear actuator with a rhombus-type flexure hinge mechanism by means of parasitic motion. Review of Scientific Instruments. 2019;90:096102. DOI: 10.1063/1.5082856

[13] Zhang X, Yu Y, Gao Q, Qiao G, Li Z, Lu X, Cheng T. A stick-slip linear piezoelectric actuator with mode conversion flexure hinge driven by symmetrical waveform. Smart Materials and Structures. 2020;29:055035. DOI: 10.1088/1361-665X/ab7f42

[14] Lu X, Gao Q, Gao Q, Yu Y, Zhang X, Qiao G, Zhao H, Cheng T. Design, modeling, and performance of a bidirectional stick-slip piezoelectric actuator with coupled asymmetrical flexure hinge mechanisms. Journal

of Intelligent Material Systems and Structures, 2020;31:1961-1972. DOI: 10.1177/1045389X20942325

[15] Wang L, Chen D, Cheng T, He P, Lu X, Zhao H. A friction regulation hybrid driving method for backward motion restraint of the smooth impact drive mechanism. Smart Materials and Structures. 2016;25:085033. DOI: 10.1088/0964-1726/25/8/085033

[16] Cheng T, Li H, He M, Zhao X, Lu X, Gao H. Investigation on driving characteristics of a piezoelectric stick-slip actuator based on resonant/off-resonant hybrid excitation. Smart Materials and Structures. 2017;26:035042. DOI: 10.1088/1361-665X/aa5c2c

[17] Cheng T, Lu X, Zhao H, Chen D, He P, Wang L, Zhao X. Performance improvement of smooth impact drive mechanism at low voltage utilizing ultrasonic friction reduction. Review of Scientific Instruments. 2016;87:085007. DOI: 10.1063/1.4960392

[18] Li H, Li Y, Cheng T, Lu X, Zhao H, Gao H. A Symmetrical Hybrid Driving Waveform for a Linear Piezoelectric Stick-Slip Actuator. IEEE Access. 2017;5:16885-16894. DOI: 10.1109/access.2017.2744498

[19] Yu Y, Gao Q, Qiao G, Zhang X, Lu X, Cheng T. A direction-guidance hybrid excitation method for inertial flexure hinge piezoelectric actuator with high speed performance. Sensors and Actuators A: Physical. 2020;314:112229. DOI: 10.1016/j.sna.2020.11222

Section 3

Modeling and Control

Chapter 5

Modeling of Piezoceramic Actuators for Control

Joel Shields and Edward Konefat

Abstract

In this chapter a full electromechanical model of piezoceramic actuators is presented. This model allows for easy integration of the piezo stack with a structural finite element model (FEM) and includes the flow of energy into and out of the piezo element, which is governed by the transducer constant of the piezo element. Modeling of the piezo stack capacitive hysteresis is achieved using backlash basis functions. The piezo model can also be used to predict the current usage of the PZT which depends on the slew rate of the voltage applied to the PZT. Data from laboratory experiments using a load frame and free response tests is used to estimate the PZT model parameters. In addition, a simplified model of a modulated full bridge strain gauge is derived based on test data which includes the effect of intrinsic bridge imbalance. Sensors of this type are often used with feedback control to linearize the behavior of the device. Taken together, the actuator and sensor model can be used for the development of piezo actuated control algorithms.

Keywords: PZT, transducer constant, strain gauge, hysteresis, charge feedback

1. Introduction

Lead zirconium titanate (PZT) was first developed around 1952 at the Tokyo Institute of Technology. Due to its physical strength, chemical inertness, tailorability and low manufacturing costs, it is one of the most commonly used piezo ceramics. It is used in various applications as an actuator including micromanipulation and ultrasonics. Due to the piezoelectric effect [1], it can also be used as a sensor of force or displacement or for energy harvesting since current is produced by the device in response to an applied strain. Three main issues, however, compromise the utility of these devices. The first is that the stroke of these devices is limited to about 40 μm, which for some applications, necessitates the use of dual stage actuation [2, 3] or inchworm actuators [4] which combine brake PZTs with actuation PZTs to extend the dynamic range of these devices. In addition, PZTs suffer from significant amounts of hysteresis (15 percent of full stroke) and creep which causes drift of the device over time. These two issues make open loop operation of these devices problematic since multiple displacements are possible for the same input and due to the fact that creep dissipates at very slow rates. To overcome these problems, it is possible to invert the hysteresis nonlinearity [5] but this can only improve accuracy by one order of magnitude and leaves the creep issue unresolved. This has led many to employ servo control with displacement sensing of the PZT elongation [6–8]. This will linearize the device to the accuracy of

IntechOpen

the sensor, and address creep, but the hysteresis nonlinearity has an impact on the loop stability which must be analyzed. In a linear sense, the hysteresis has an effect on the both the loop gain and phase margin. These impacts necessitate high fidelity modeling of the PZT behavior, which can be quite rich and faceted. The model described in this chapter is intended to be used as a necessary precursor for control design and simulation.

Many models for PZT stacks have been developed in the literature, such as those in [8–10], but few incorporate the transfer of energy into a structure and out of the structure back into the device. Of those that do describe this important detail, [1, 11–13], the treatment of piezoelectric hysteresis in the model is often problematic in that either friction elements, such as the Maxwell resistive capacitor (MRC) are used, which are difficult to simulate, or nonlinear differential equations are used which have very heuristic fitting functions [1]. The hysteresis model described here overcomes these difficulties by using backlash basis functions which are relatively easy to fit to experimental data and are numerically efficient to simulate. The hysteresis model does not included creep, but it is possible to add springs and dashpots to the backlash elements which will capture the creep behavior.

The PZT model presented in this chapter can also be used to predict current usage which is important in terms of amplifier design and slew rate limitations. It can also be easily integrated into a FEM of the structure that the PZT is embedded in. The model in this chapter is based on the work in [1, 12]. Unfortunately, these papers offer little in terms of determining the parameters of the model. In particular, the internal capacitances and transducer constant vary from one type of PZT to another. As a result, this chapter describes experimental techniques for determining these parameters regardless of the particular PZT configuration. We also validate proper operation of our test equipment and compare the simulated model output to test data.

For control design purposes it is also important to model the sensor in addition to the actuator. Various types of sensors have been used for PZT control such as optical encoders with 1.2 nm resolution, laser metrology gauges, and more commonly, strain gauges. Strain gauges offer good performance for the price and can be directly bonded to the sides of the PZT. A model of a modulated full strain gauge bridge is given which includes the effect of intrinsic bridge imbalance and bridge sensitivity. A simplified version of this model, excluding all exogenous signals, is given to facilitate loop shaping of a control loop.

PZTs are often used in optical instrument such as coronagraphs, interferometers, spectrometers and microscopes where micro-manipulation below the scale of optical wavelengths is required. Given the small wavelengths of visible and infrared light (350 nm - 14 μm), subnanometer positioning is required for these applications. PZTs come in a wide variety of shapes and configurations. Here we focus on piezo stack actuators which are the most commonly used. In particular, we tested and modeled the P-888.51 PICMA stack multilayer piezo actuator [14] which is ubiquitous in its use for various missions at JPL. This PZT has dimensions of 10 mm $\times$ 10 mm $\times$ 18 mm and has a nominal stroke of 15 μm at 100 Volts across its terminals. It has a blocking force of up to 3600 Newtons and a mechanical stiffness of 200 $N/\mu m$. Nominal voltage range applied to the PZT is 0–100 Volts, but it is possible to apply between -20 - 120 Volts for extended stroke. These devices are operated with large preloads to keep the device in compression during operation. Dynamic loads that put the ceramic device in tension need to be avoided since they are brittle and fracture easily in tension. The recommended preload is 15 MPa which for the 10 mm $\times$ 10 mm cross sectional area of the device is 1500 Newtons. The first mode of this particular PZT is 70 kHz which is far greater than the first mode of any structure it is generally incorporated into.

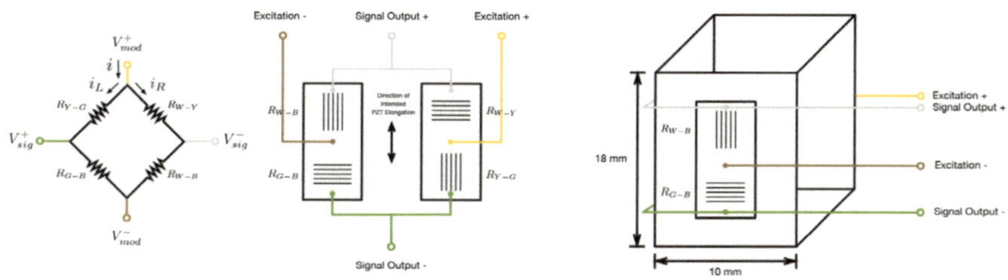

Figure 1.
From left to right, circuit diagram of strain gauge Wheatstone bridge showing resistive elements labeled by the color of wires on each side of the resistor, physical layout of resistive elements and wiring connections, and strain gauge elements bonded to the PZT.

The PICMA P-888.51 can be ordered with a full bridge strain gauge bonded directly to the sides of the PZT ceramic. As mentioned above, strain gauges are used to linearize the hysteretic behavior of PZTs and to mitigate their drift or creep, which can be substantial. To determine the bridge response, we analyze the modulating electronics used to read the Wheatstone bridge. The modulation is a technique commonly used for rejection of noise picked up along the cabling between the PZT and electronics board. The strain gauge bridge we analyze is composed of two Vishay half bridges (Part number N2A-XX-S053P-350) bonded to opposite sides of the PZT as shown in **Figure 1**. Each bridge resistor has a nominal resistance of 350 Ohms.

Three identical PZTs were tested which we refer to by their serial numbers, SN637, SN629 and SN618.

2. Full electro-mechanical PZT model

The PZT model we analyze is depicted in **Figure 2**. It is taken from the model proposed in [1, 12, 13] with a few modifications. We have added a resistor on the input since this is part of the amplifier used to drive the device. We depict the resistor in **Figure 2** as having a switch since most of our experiments were conducted without this resistor. This resistor and the equivalent capacitance of the circuit determines the RC time constant of the actuator model. This time constant can vary based on the voltage input (V_{pea}) history due to the nonlinear capacitance, $C_h(\cdot)$. As current, $\dot{q}$, flows into the circuit charge develops across the working

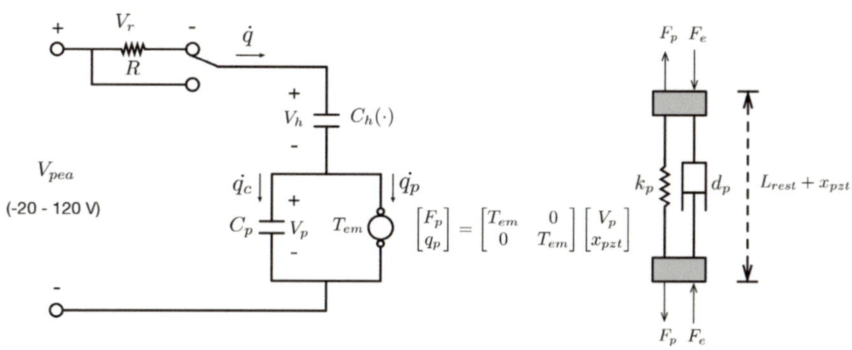

Figure 2.
Nonlinear electro-mechanical PZT model. Features of this model include the nonlinear capacitance, $C_h(\cdot)$, and transducer constant, T_{em}, which governs the flow of energy into and out of the PZT.

capacitor, C_p, which due to the *inverse piezoelectric effect* produces a force, F_p in the piezoelectric stack. Conversely, any environmental force, F_e, which causes velocity of the PZT stack generates a current, $\dot{q}_p$, in the circuit due to the *piezoelectric effect*.

The unknown model parameters in **Figure 2** include the two capacitances, C_p and $C_h(\cdot)$, and the transducer constant, T_{em}. The resistor value, R, is known by design (80 Ohms) and can be easily measured. The mechanical stiffness of the PZT, k_p, can be obtained from Physik Instrumente (PI) spec. Sheets [14]. The damping coefficient, d_p, of the PZT ceramic is a structural parameter of the PZT and not part of this investigation. It is primarily a parameter that effects the high frequency behavior of the PZT in terms of changing the modal response. Here we are more concerned with the low frequency, nonlinear behavior of the device.

Writing some simple circuit equations based on **Figure 2**, we have Kirchhoff's Voltage Law (KVL) and Kirchhoff's Current Law (KCL),

$$V_{pea} = V_r + V_h + V_p \qquad (KVL) \tag{1}$$

$$\dot{q} = \dot{q}_c + \dot{q}_p \qquad (KCL). \tag{2}$$

We also have the three constitutive relations for each of the components in this circuit,

$$q = C_h(V_h) \qquad q_c = C_p V_p \tag{3}$$

$$V_r = i_r R \qquad with \qquad q(t) = \int_0^t i_r(t)dt. \tag{4}$$

Using Eqs. (1)–(4) we can convert the circuit diagram in **Figure 2** to the block diagram in **Figure 3**. **Figure 3** is easier to simulate since the mixed electrical and mechanical domains are irrelevant. Few simulation software applications allow for mixed domain signals. SPICE is a typical example of this. Simscape™ does allow for mixed domain signals but nonlinear multivalued capacitors are not supported. Putting the model in block diagram form allows for easy integration in the Matlab® and Simulink® programming environment which supports simulation of both a structural model for the mechanical impedance, $M(\cdot)$, and hysteretic capacitance, $C_h(\cdot)$, using primitive Simulink components. Looking at **Figure 3** KVL is represented with the first two summing nodes. The center summing node is depicting an integrated version of KCL, Eq. (2). Since the voltage across the resistor, V_r, is proportional to the current, $\dot{q}$, the charge, q, needs to be differentiated, thus we use the Laplace operator, s, in **Figure 3** to represent this. The saturation block on

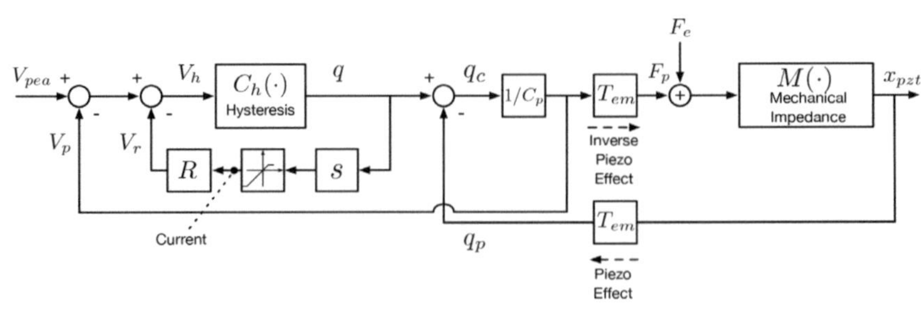

Figure 3.
Block diagram of PZT model showing the inverse piezo effect and piezo effect with charge feedback. The nonlinear capacitance function, $C_h(\cdot)$, is modeled with backlash operators. The mechanical impedance, $M(\cdot)$, can be as simple as a spring representing the mechanical stiffness of the PZT material or a FEM of the structure the PZT is embedded into.

current is included to capture current limits of the PZT amplifier. As an example of this, the flight amplifier for the Nancy Grace Roman Space Telescope has a limit of +2 mA for sourcing of current into the PZT and − 14 mA for sinking of current out of the PZT.

At steady state and with the boundary conditions at both ends of the PZT free (i.e. no structural stiffness to oppose the PZT produced force, F_p.), the mechanical impedance of the PZT, $M(\cdot)$, reduces to $1/k_p$. The displacement of the PZT can then be written as,

$$x_{pzt} = \frac{1}{k_p}(F_e + F_p), \tag{5}$$

with $F_p = T_{em}V_p$. With the terminals of the PZT open circuit $V_p = V_{pea}$ assuming the device has been shorted and drained of all charge prior to applying an environmental force. Since V_{pea} is a measurable signal this allows us to solve for T_{em} using Eq. (5),

$$T_{em} = \frac{k_p x_{pzt} - F_e}{V_{pea}}, \tag{6}$$

where both x_{pzt} and F_e can also be measured. This equation forms the basis of the experiments described in Section 3.2.

To determine the nonlinear capacitance, $C_h(\cdot)$, and working capacitance, C_p, we collect input/output data while applying a voltage, V_{pea}, to the PZT with the environmental force, F_e, set to zero. To determine the hysteresis of this block over its full range of inputs, we apply a sinusoidal V_{pea} voltage that goes from 0–100 Volts. This voltage is applied slowly so that we can assume that $M(\cdot) \approx 1/k_p$. The output signal of $C_h(\cdot)$, q, can be obtained by integrating measurements of the current flowing into the PZT as a result of the voltage, V_{pea}, applied to the PZT terminals. The input signal to $C_h(\cdot)$, V_h, can be solved for using KVL,

$$V_h = V_{pea} - V_r - V_p, \tag{7}$$

where $V_r = 0$ since during this test the resistor R is not used. V_p in Eq. (7) can be solved for by using Eq. (5) with $F_e = 0$ and $F_p = T_{em}V_p$. This leads to,

$$V_p = \frac{x_{pzt}k_p}{T_{em}} \tag{8}$$

Eqs. (7) and (8) together with the measured charge, q, form the basis of the experiments described in Section 3.3 for determining the nonlinear capacitance, $C_h(\cdot)$.

From the block diagram in **Figure 3** we have,

$$V_p = \frac{1}{C_p}(q - T_{em}x_{pzt}). \tag{9}$$

Plugging in $F_e = 0$ and $F_p = T_{em}V_p$ with V_p given by Eq. (9) into Eq. (5) we have,

$$x_{pzt} = \underbrace{\frac{T_{em}}{k_p C_p + T_{em}^2}}_{k_{q2x}} q. \tag{10}$$

We can solve for the gain, k_{q2x}, in a least squares sense,

$$k_{q2x} = \left(\mathbf{q}^T\mathbf{q}\right)^{-1}\mathbf{q}^T\mathbf{x}_{pzt} \tag{11}$$

where $\mathbf{q}$ and $\mathbf{x}_{pzt}$ are vectors of the sampled signals q and x_{pzt}. The working capacitance, C_p, is then,

$$C_p = \frac{T_{em} - T_{em}^2 k_{q2x}}{k_p k_{q2x}} \tag{12}$$

Eqs. (10)–(12) form the basis of the experiments described in Section 3.3 for determining the working capacitance, C_p.

The strain gauge voltage change is the difference between the voltage at the start of the test and at the time that the peak force was applied.

3. Experiments

3.1 Displacement sensing for test campaign

Both experiments described below require sensing of the PZT elongation. To do this we use the strain gauge bonded to the PZT. For testing purposes we applied a DC voltage (10 Volts) as the excitation input to the bridge and configured a differential amp on the output of the bridge. The gain of this amplifier was set to 311.4 which amplified the millivolt signal of the bridge output to the 2–3 Volt range which was easily measurable by our test equipment. The amplified voltage was calibrated to the actual displacement of the PZT using a micrometer with 0.5 micron accuracy (Mitutoyo MDH high accuracy digimatic micrometer, series 293.). A linear fit of the calibration data, measured displacement vs. measured voltage, revealed a fit error of 0.71 microns for the worst case PZT with scale factors of 3.077 ($\mu m/V$), 3.183 ($\mu m/V$) and 3.246 ($\mu m/V$) for SN637, SN629 and SN618 respectively. Calibrated steady state measurements of the strain gauge had an RMS of 22 nm which was more than sufficient for our testing campaign.

3.2 Experiment 1: instron testing

As discussed above we use the force balance Eq. (6) to determine the transducer constant, T_{em}. The environmental force, F_e, in this equation was applied using an Instron 8801 Servohydraulic load frame which is part of the JPL Materials Testing and Failure Facility. Refer to **Figure 4** for a picture of the PZT test jig. To avoid potential damage to the load frame the hydraulic press was operated in displacement mode which left force an uncontrolled, but measured parameter (see **Figure 5**). To measure the back EMF term, V_{pea}, in this equation a voltmeter with high input impedance is required. Without a high impedance voltmeter the RC decay time constant would be much faster than the duration of the experiment and compromise the data. We used a Keithley 617 high resistance meter with 200 TeraOhm input impedance to keep charge from bleeding off the PZT leads. To measure the PZT displacement an Agilient 34401A DMM was used to monitor the amplified bridge output voltage. The applied compressive load to the PZT was measured with the load cell that is incorporated into the load frame.

Each of the three PZTs was tested with and without the load resistor to see how much resistive force is created by the charge feedback. Tests done with the resistor

Figure 4.
PZT installed in the Instron jig between two compression platens. The load frame is used to apply a compressive load of 800 (N) while the PZT displacement, back EMF, and applied load are measured. The PZT is compressed with and without a 200 Ohm resistor applied across the leads of the PZT. This resistor is shown on the right terminal block in the open circuit configuration.

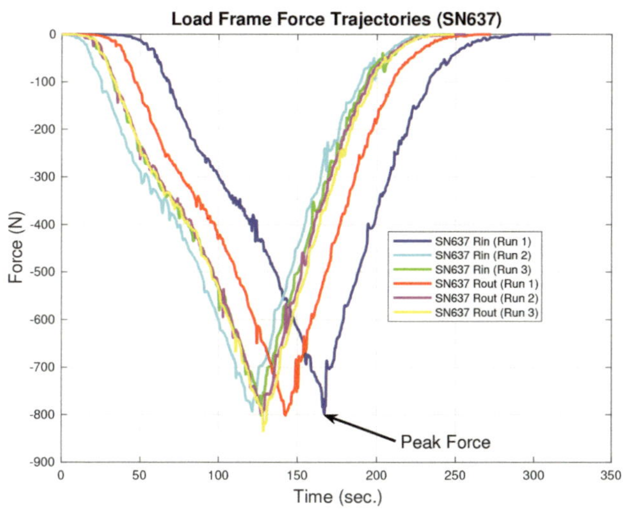

Figure 5.
PZT signal set recorded during each free response test. The charge signal was obtained by integrating the current. The position signal was obtained by sampling the strain gauge voltage and applying the calibration described in Section 3.1.

bleeds off any charge that would otherwise develop and leaves only the mechanical stiffness to oppose the applied compressive force. In addition, to verify repeatability of the experimental results each tested configuration was repeated three times. **Table 1** summarizes the results of these tests. Note that with the load resistor the peak displacements are substantially greater than the cases without the load resistor even though the peak loads applied are similar. This is the result of charge feedback generating an opposing force beyond that provided by the mechanical stiffness alone. It may be surprising to find that the electrical force produced is close to or even greater than the force produced by the mechanical stiffness when the resistor is not used. Plugging in the peak force, Fe, the peak displacement, x_{pzt}, and peak back EMF, V_{pea}, from **Table 1** into Eq. (6), and using the spec. Sheet value [14] of the mechanical stiffness, $k_p = 200 \ (N/\mu m)$, we can solve for the transducer

R In	SN637	SN637	SN637	SN629	SN629	SN629	SN618	SN618	SN618
	(Run 1)	(Run 2)	(Run 3)	(Run 1)	(Run 2)	(Run 3)	(Run 1)	(Run 2)	(Run 3)
Strain Gauge Voltage Change (V)	−0.925	−1.040	−0.958	−1.112	−0.918	−0.944	−0.885	−0.914	−0.926
Peak Displacement (um)	**−2.847**	**−3.200**	**−2.949**	**−3.538**	**−2.922**	**−3.006**	**−2.873**	**−2.966**	**−3.005**
Peak Force (N)	−772.2	−778.1	−774.3	−799.5	−738.6	−773.3	−792.8	−798.8	−809.8
Peak Back EMF (V)	0.0	0.0	0.0	0.0	0.0	0.0	0.0	0.0	0.0
R Out	**SN637**	**SN637**	**SN637**	**SN629**	**SN629**	**SN629**	**SN618**	**SN618**	**SN618**
	(Run 1)	(Run 2)	(Run 3)	(Run 1)	(Run 2)	(Run 3)	(Run 1)	(Run 2)	(Run 3)
Strain Gauge Voltage Change (V)	−0.415	−0.413	−0.426	−0.382	−0.381	−0.384	−0.334	−0.334	−1.078
Peak Displacement (um)	**−1.277**	**−1.272**	**−1.311**	**−1.217**	**−1.213**	**−1.221**	**−1.084**	**−1.084**	**−0.336**
Peak Force (N)	−798.8	−797.1	−834.3	−807.2	−795.8	−800.0	−808.8	−815.4	−844.0
Peak Back EMF (V)	11.68	12.32	12.55	11.98	12.51	12.63	12.20	12.21	12.15

Table 1.
Instron testing data with and without load resistor. For each PZT, SN637, SN629 and SN618 the load test was repeated three times.

constant, T_{em}. The mean value of T_{em} over each PZT and over the three runs for each PZT was 46.71 (N/V) or (C/m). The variance of this estimate was minimal.

3.3 Experiment 2: free-free test

To perform the free-free test on the PZT, the applied voltage, resulting current applied to the PZT and PZT displacement all need to be measured synchronously. To do this we used a Keysight Technologies B2902A precision source to sample the applied voltage and current. This instrument was designed to measure I/V measurements easily and accurately. To measure the PZT displacement the output of the bridge differential amp was measured with a Tektronics oscilloscope. To synchronize the two data sets we used the peak value in each data record. These signals are shown in **Figure 5** together with the applied charge which is just the integrated current. Before each test a relay was used to temporarily short the leads of the PZT to drain any charge. This put the PZT in its rest state prior to each test.

To validate our instrumentation we compared the current and charge reported by our test setup using a simple 5.5 μF capacitor against a SPICE simulation with the same capacitor value. The peak charges agreed to within 2 percent providing confidence to the current measurements.

Using the PZT position and charge measurements shown in **Figure 6** along with the previously estimated value of the transducer constant, T_{em}, and stiffness, k_p, taken from the PI spec. Sheet [14], we can use Eqs. (10)–(12) to solve for the working capacitance, C_p. This resulted in a value of 5.802 μF for SN637 which is close to the small signal (1 V_{pp}) effective capacitance value listed in the PI spec. Sheet of 6.0 $\mu F \pm 20\%$. Note the effective capacitance in our model is the series

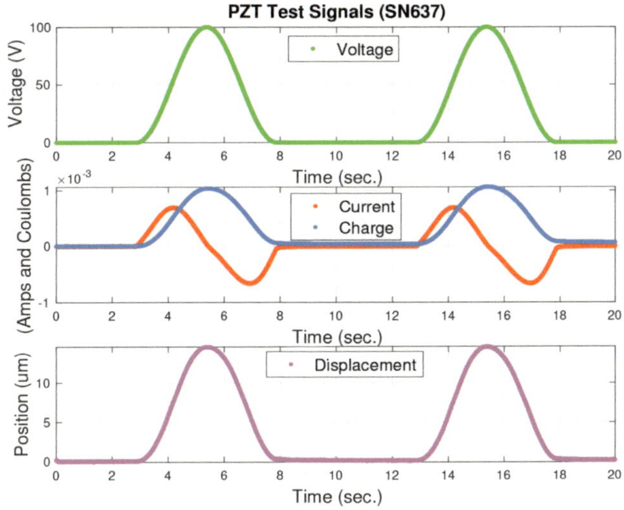

Figure 6.
Force trajectories during operation of the load frame in displacement mode. Each of the three PZTs was tested three times with the load resistor, R_{in} cases, and three times without the load resistor, R_{out} cases. Since force was not a controlled variable during the test it had a noticeable but tolerable amount of variation or jitter.

combination of C_p with $C_h(\cdot)$ which is $C_pC_h(\cdot)/(C_p + C_h(\cdot))$. Thus, whatever the value of $C_h(\cdot)$ it can only act to reduce the effective capacitance from the baseline value of C_p.

Before determining the nonlinear function, $C_h(\cdot)$, the input signal, V_h, is generated using Eqs. (7) and (8). This revealed that the voltage drop across the hysteretic capacitance, V_h, was approximately one-third of the applied voltage, V_{pea}. This leaves two-thirds of the applied voltage across the working capacitor to produce work. The output of the nonlinear function, $C_h(\cdot)$, is given by the charge measurements, q. The input and output data is shown in **Figure 6** which demonstrates a great deal of hysteresis. Note that creep between the two applied sinusoids infects the data as can be seen in the lower left of **Figure** 7. This portion of the data was not fit by the model described below.

3.3.1 Hysteresis model

To fit the capacitive hysteresis data a neural network of backlash basis functions is employed of the form,

$$q = \sum_{i=1}^{N} B_i\left(V_h, x_i^o, w_i\right) \cdot k_i \tag{13}$$

where $B_i\left(V_h, x_i^o, w_i\right)$ is a backlash operator with input variable, V_h, initial condition, x_i^o, and deadband width, w_i. Each backlash output is weighted with a gain k_i which changes the slope of the backlash operator when the deadband is engaged. Details of fitting the parameters of this model to experimental data are given in [5, 15] but, to summarize, the gains k_i are used to fit the curvature of the hysteresis data, the schedule of deadband widths, w_i, are used to accurately capture regions of the hysteresis that have high curvature and the initial conditions are used to capture the uniqueness of the initial loading curve. A diagram of the hysteresis model is provided in **Figure 8**.

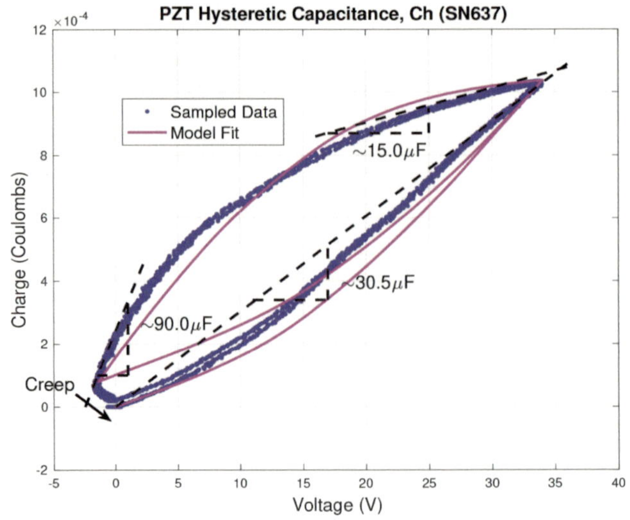

Figure 7.
Capacitive hysteresis data and model fit. Linearized approximations are shown with the large signal approximation of ~30.5 μF.

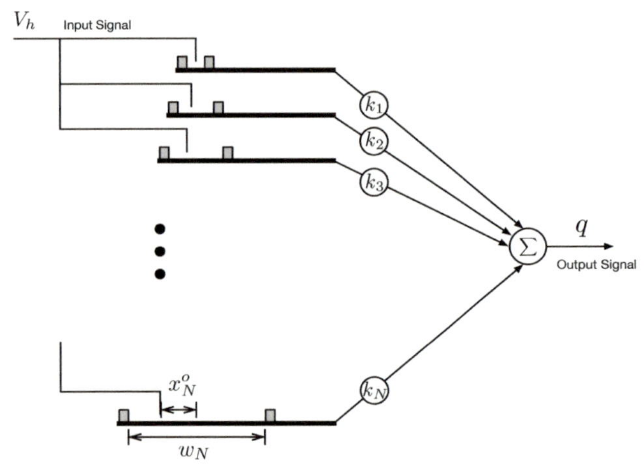

Figure 8.
Diagram of capacitive hysteresis model used to fit the data in ***Figure*** *7. Initial state and deadband width are indicated for the Nth element of the model. The output arm displacement of each backlash element is multiplied by gain k_i and summed with the other backlash elements to form the output variable q.*

4. Complete model

To validate the PZT model described in this chapter we exercise the model in **Figure 3** by applying a full scale (0–100 Volt) sinusoidal input voltage, V_{pea}, and record the model output, x_{pzt}. This data can then be compared with the same data acquired during the experiments described in Section 3.3. This comparison is made in **Figure 9** which shows good agreement between the experimental data and model data. In addition, comparisons of the experimentally measured current and modeled current signal also demonstrated similar agreement. Note the hysteresis percentage in **Figure 9**, as defined by the vertical thickness of the hysteresis loops relative to the peak displacement, is lessened for the output signal of the model relative to the percentage of hysteresis demonstrated in **Figure 7**. This is most likely

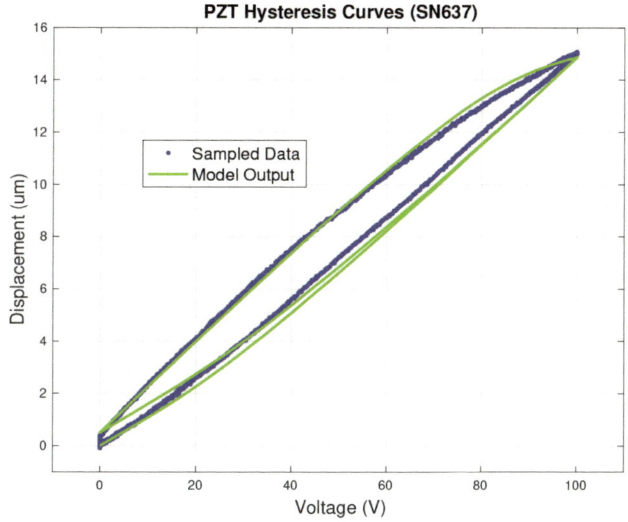

Figure 9.
Experimental and modeled input/output data for the full PZT model.

due to the voltage feedback terms, V_p and V_r, in **Figure 3** that act to reduce the uncertainty caused by the capacitive hysteresis, $C_h(\cdot)$.

5. Modulated strain gauge model

We have seen from the test data that the behavior of the PZT is nonlinear due to its hysteresis. Many applications of PZT actuators require greater positioning accuracy than what can be achieved with open loop operation of the PZT. To overcome this accuracy issue, it is common to use strain gauges bonded directly to the face of the PZT. These sensors are used with feedback control to greatly improve the positioning accuracy of the actuator. Typically the accuracy can be improved from 12–15 percent down to 0.1 percent by using strain gauge sensors. Strain gauge sensors also suffer from hysteresis (This is what limits their accuracy.) but the amount of hysteresis is much less than the actuator.

Before designing any control loop that uses strain gauge feedback, it is necessary to understand the output response of the strain gauge voltage versus the input strain. To do this, we first characterize the resistive changes of the four Wheatstone bridge resistors as a function of the PZT elongation. The results of this analysis can then be used to derive the sensitivity of the bridge output. This analysis assumes that the strain gauge bridge is modulated with a square wave reference at its excitation terminals. The modulation is done in order to reduce noise pickup in the cabling between the bridge location and PZT processing electronics board which can be separated by several meters in typical applications. Any noise with a frequency content well below the modulation frequency of 10 kHz can be eliminated with this technique. Demodulation can be done in analog electronics but here we describe a case where the demodulation is done in FPGA firmware.

5.1 Bridge resistance changes

The stain gauge configuration we study is a full Wheatstone bridge, meaning that all four resistive elements of the bridge see a change in strain due to the elongation of the PZT. Two of the bridge elements, on opposite diagonals of the

Wheatstone bridge, are bonded to the PZT along the length of the PZT which is the direction of intended length change. The remaining two bridge elements are bonded to the PZT perpendicular to this, along the width of the actuator. These two bridge elements will also see a resistance change when the PZT elongates due to the fact that as the PZT lengthens it also gets more narrow. The circuit diagram of the Wheatstone bridge and layout of the bridge elements when bonded to the PZT is shown in **Figure 1**.

To characterize the exact relationship between the PZT elongation and resistive change of each bridge element we conducted a test where the resistance measurement across each strain gauge resistor was measured for several increments of PZT displacement from its rest length to its maximum displacement. Each resistance measurement is the equivalent resistance of the element across the probe terminals in parallel with the other three elements, i.e.,

$$R_{m1} = R_{W-Y} || R_{Y-G} + R_{G-B} + R_{W-B} \tag{14}$$

$$R_{m2} = R_{W-B} || R_{W-Y} + R_{Y-G} + R_{G-B} \tag{15}$$

$$R_{m3} = R_{G-B} || R_{W-B} + R_{W-Y} + R_{Y-G} \tag{16}$$

$$R_{m4} = R_{Y-G} || R_{G-B} + R_{W-B} + R_{W-Y}. \tag{17}$$

These measurements must be solved for the individual bridge elements by solving the system of equations in (14)–(17) for each of the four bridge element resistances. This system of equations is nonlinear but can be solved uniquely using a nonlinear equation solver. Mathematica was used for this purpose. Solving this system of equations resulted in three complex solutions and one real solution for each of the resistances on the right hand side. Since the resistance must be real, the real solution was taken as the answer.

The results of these measurements are shown in **Figure 10** for PZT SN618. The two strain gauges that are oriented along the length of the PZT increase their resistance as the PZT elongates since this deformation lengthens the gauge and thins the wires of the bridge. The two strain gauges oriented along the width of the PZT

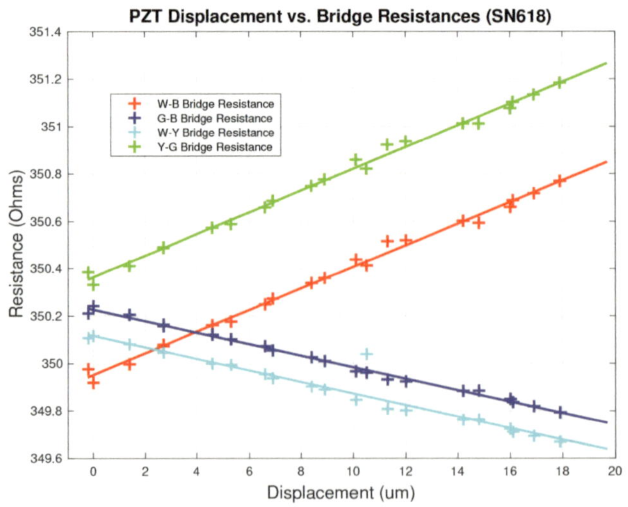

Figure 10.
Four bridge resistances as a function of PZT displacement. This data is from 1 of 3 PZTs tested. Note the difference in slopes of the four resistors, one negative slope for the two horizontally placed bridge elements and one positive slope for the two vertically placed bridge elements. Initial resistances at zero elongation reveals the intrinsic bridge imbalance due to manufacturing tolerances.

decrease in resistance since, in this direction, the PZT narrows. As it narrows the wires of these bridge resistors become shorter and thicker reducing their resistance. Interestingly, the ratio of the slopes of the horizontally mounted gauges to the vertically mounted gauges is equal to the aspect ratio of the PZT, W_{rest}/L_{rest}, where W_{rest} is the rest width of the PZT and L_{rest} is the rest length of the PZT. Also note that at zero elongation of the PZT, the actual resistance of each bridge element deviates from the nominal value of 350 Ohms. For all three full bridge strain gauges tested the variation from the nominal resistance for each bridge element was found to be 0.224 Ohms, 1-σ with a maximum error bounded by ± 0.4 Ohms. This was consistent with the spec. Sheet tolerance [16].

Taking into account the nominal resistance, variation in nominal resistance and strain induced resistance change, the resistance of each bridge element can be written as,

$$R_{W-Y} = \left(R_{nom} + \Delta R_{nom}^{W-Y}\right) - k_{gf} \cdot \left(R_{nom} + \Delta R_{nom}^{W-Y}\right) \cdot \left(\frac{1}{L_{rest}}\right) \cdot \left(\frac{W_{rest}}{L_{rest}}\right) \cdot s_{pzt} \tag{18}$$

$$R_{G-B} = \left(R_{nom} + \Delta R_{nom}^{G-B}\right) - k_{gf} \cdot \left(R_{nom} + \Delta R_{nom}^{G-B}\right) \cdot \left(\frac{1}{L_{rest}}\right) \cdot \left(\frac{W_{rest}}{L_{rest}}\right) \cdot s_{pzt} \tag{19}$$

$$R_{Y-G} = \left(R_{nom} + \Delta R_{nom}^{Y-G}\right) + k_{gf} \cdot \left(R_{nom} + \Delta R_{nom}^{Y-G}\right) \cdot \left(\frac{1}{L_{rest}}\right) \cdot s_{pzt} \tag{20}$$

$$R_{W-B} = \left(R_{nom} + \Delta R_{nom}^{W-B}\right) + k_{gf} \cdot \left(R_{nom} + \Delta R_{nom}^{W-B}\right) \cdot \left(\frac{1}{L_{rest}}\right) \cdot s_{pzt}, \tag{21}$$

where k_{gf} is the *gauge factor* of the bridge element ($k_{gf} = 2.05$ [11]) which relates the change in resistance to the strain, (s_{pzt}/L_{rest}). R_{nom} is the nominal bridge resistance and $\Delta R_{nom}^{W-Y}, \Delta R_{nom}^{G-B}, \Delta R_{nom}^{Y-G}, \Delta R_{nom}^{W-B}$ are the variations from this nominal value. In the following treatment of bridge sensitivity, these intrinsic imbalance terms can be ignored since they are a small portion of the total resistance.

5.2 Bridge sensitivity

Referring to **Figure 9** we can derive the bridge sensitivity using simple circuit analysis. Writing out the constitutive relation for each resistor we have,

$$V_{mod}^{+} - V_{sig}^{+} = i_L \cdot R_{Y-G} \tag{22}$$

$$V_{mod}^{+} - V_{sig}^{-} = i_R \cdot R_{W-Y} \tag{23}$$

$$V_{sig}^{+} - V_{mod}^{-} = i_L \cdot R_{G-B} \tag{24}$$

$$V_{sig}^{-} - V_{mod}^{-} = i_L \cdot R_{Y-G}, \tag{25}$$

where i_L and i_R are the currents indicated in **Figure 1**. Solving this system of equations for i_L, i_R, V_{sig}^{+} and V_{sig}^{-} as functions of the two excitation voltages V_{mod}^{+} and V_{mod}^{-} gives the output voltage, $V_{out} = V_{sig}^{+} - V_{sig}^{-}$, as,

$$\begin{aligned} V_{out} = V_{sig}^{+} - V_{sig}^{-} &= \frac{R_{W-B}}{R_{W-B} + R_{W-Y}} V_{mod}^{+} + \frac{R_{W-Y}}{R_{W-B} + R_{W-Y}} V_{mod}^{-} \\ &- \frac{R_{G-B}}{R_{G-B} + R_{Y-G}} V_{mod}^{+} - \frac{R_{Y-G}}{R_{G-B} + R_{Y-G}} V_{mod}^{-}. \end{aligned} \tag{26}$$

Substituting the resistances, Eqs. (18)–(21), into Eq. (26) and letting $V^+_{mod} = V_{mod}$ and $V^-_{mod} = -V_{mod}$ we can express the output voltage of the bridge as a function of the displacement x_{pzt},

$$V_{out} = f(s_{pzt}) = \frac{2V_{mod}k_{gf}(L_{rest} + W_{rest})x_{pzt}}{(2L^2_{rest} + k_{gf}L_{rest}x_{pzt} - W_{rest}k_{gf}x_{pzt})}. \tag{27}$$

Note that the output voltage of the Wheatstone bridge is actually weakly nonlinear in the PZT displacement, x_{pzt}, which is a bit surprising. Also note that the signal level of the output is directly proportional to the modulation voltage, V_{mod}. This implies that the resolution of the device can be arbitrarily increased with larger excitation voltages. The cost of this increased resolution would be an increased thermal signature. The thermal signature can be an issue in some applications since it can cause excessive drift of the electronics and warping of any optics near the bridge.

To derive the linearized gain of Eq. (27), we can use the first term of a Taylor series,

$$\frac{V_{out} - f(x_{pzt_o})}{x_{pzt} - x_{pzt_o}} = \left.\frac{\partial f}{\partial x_{pzt}}\right|_{x_{pzt}=x_{pzt_o}} = \left.\frac{(4L^2_{rest}V_{mod}k_{gf}x_{pzt}(L_{rest} + W_{rest}))}{(2L^2_{rest} + k_{gf}L_{rest}x_{pzt} - k_{gf}W_{rest}x_{pzt})^2}\right|_{x_{pzt}=x_{pzt_o}}. \tag{28}$$

Evaluating this derivative at $x_{pzt_o} = 7.5(\mu m)$ gives a gain of 354.18 (V/m) between the change in PZT elongation from x_{pzt_o} and the change in differential bridge output voltage, $V_{out} - f(x_{pzt_o})$.

5.3 Simplified model

After the bridge, a differential amplifier is used to boost the voltage to the ±2.5 Volt range of the A2D converter (COBHAM RAD1419 with 14 bits). This sampling is done at high rate, 800 kHz, to capture the high and low modulation levels of the 10 kHz square wave. This gives 40 samples during each portion of the modulation signal. These samples are averaged and then differenced before being operated on in firmware. The FPGA firmware applies a linear transformation to this signal in order to map it to the desired DAC range of ±2.6 Volts for the feedback servo. This bias

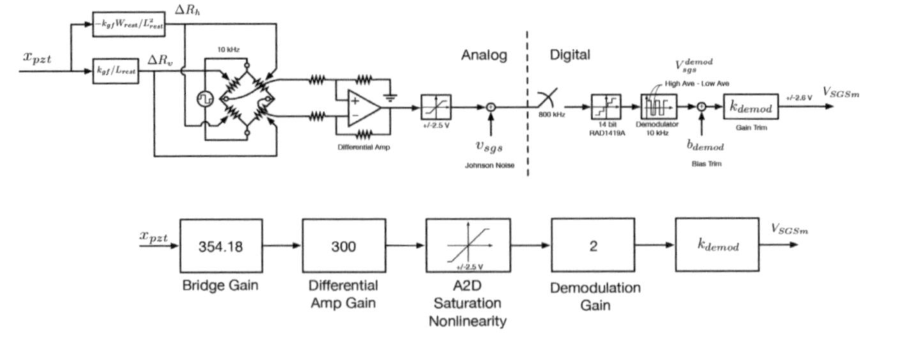

Figure 11.
Circuit schematic (top subfigure) and simplified block diagram of the strain gauge processing electronics (bottom subfigure). Modulation of the Wheatstone bridge is done with a 10 kHz square wave with 2 Volts applied to the reference excitation terminal and − 2 Volts applied to the reference ground terminal. These voltages are swapped every 50 usec. The bridge voltages are sampled at 800 kHz and demodulated in firmware.

and scale factor trimming can be done in analog electronics but is more accurately done in firmware. The full signal chain from PZT displacement to demodulated firmware voltage is shown in the top subfigure of **Figure 11**. The gain of the differential amp cannot map its output voltage exactly to the rails of the A2D since the intrinsic imbalance of the gauge and resulting bias voltage at zero PZT displacement prevents this.

For purposes of control loop design a simplified model of the strain gauge response is useful. If we ignore exogenous signals, such as the gauge noise and bias trim, which do not effect the loop gain, a simplified linear model of gauge response is shown in the bottom subfigure of **Figure 11**. The bridge gain, the differential amp gain of 300, demodulation gain of 2 and trim scale factor are all indicated in this figure. The demodulation gain is 2 since the firmware takes the difference between the same two voltages with opposite sign. Eqs. (27) and (28) give the sensitivity of each of these voltage levels and not their difference.

6. Conclusions

In this work we have developed a full electro-mechanical model of piezoelectric actuators and determined the parameters of this model. Unique experiments were designed to determine the transducer constant and capacitances of the model. The hysteretic capacitance was fit with backlash basis functions which was proposed by the first author in [15]. This hysteresis model is numerically efficient and captures the multivalued behavior of hysteresis as well as the curvature of the hysteresis loops. The output of the PZT model agreed well with the experimental data and successfully predicts the current draw of the actuator which is an important feature of the model for comparison against power limits and slew rate requirements.

For control design actuator models are only half of what is required. Models of the sensors used is also important. In this work we focused on the use of strain gauge sensors which are commonly used with PZT actuators. A nonlinear model of the strain gauge full bridge was developed from which a linearized model was generated. This linear model included the effects of Wheatstone bridge sensitivity, differential amplification, demodulation and firmware scaling.

Although not the focus of this work, the PZT model that we have developed and experimentally identified could easily be included into a FEM of the structure that the PZT is intended to move. This has been done for the fast steering mirror (FSM) used by the Nancy Grace Roman Space Telescope. This FSM has three PZTs that are used to actuate special flexures that amplify the PZT elongation. This amplified motion is used to move a mirror flat in tip, tilt and piston.

One issue that is often over looked with piezo devices is the creep that is produced by these devices. This makes open loop operation with these actuators very challenging. With sensing the creep is usually slow enough to be effectively cancelled by using feedback. Nonetheless, a full piezo model with creep has not been successfully developed to the knowledge of the authors. Augmentations to the backlash elements presented in this work have shown promise in this area but further investigations are necessary.

Acknowledgements

The work described in this paper was carried out at the Jet Propulsion Laboratory, California Institute of Technology, under contract with the National

Aeronautics and Space Administration. The author wishes to thank the Nancy Grace Roman Space Telescope Project for funding this work.

Conflict of interest

The authors declares no conflict of interest.

Author details

Joel Shields* and Edward Konefat
Jet Propulsion Laboratory, California Institute Technology, Pasadena, CA, USA

*Address all correspondence to: Joel.f.shields@jpl.nasa.gov

IntechOpen

© 2021 The Author(s). Licensee IntechOpen. This chapter is distributed under the terms of the Creative Commons Attribution License (http://creativecommons.org/licenses/by/3.0), which permits unrestricted use, distribution, and reproduction in any medium, provided the original work is properly cited.

References

[1] Adriens H, De Koning WL, Banning R. Modeling of Piezoelectric Actuators. IEEE Transactions on Mechantronics. 2000;**5**(4):331-341

[2] Lurie B, Hench J, Ahmed A, Hadaegh F. Nonlinear Control of the Optical Delay Line Pathlength. Orlando: SPIE AeroSense; 1999. pp. 1-11

[3] Atsumi T, Nakamura S, Furukawa M, Naniwa I, Xu J. Triple-Stage-Actuator System of Head-Positioning Control in Hard Disk Drives. IEEE Transactions on Magnetics. 2013;**49**(6):2738-2743

[4] Shields J. Asynchronous Control of a Prototype Inchworm Actuator: Control Design and Test Results. Actuators. 2019;p. 1–21.

[5] Shields J. Characterization and Inversion of the AMD Primary and Secondary Rigid Body Actuators. Pasadena, (CA): Jet Propulsion Laboratory; 2010. D-60616.

[6] Yan Z, Sun L. Design, Control and Application of a PZT-driven Micro-stage. In: International Conference on Mechatronics and Automation. IEEE: Harbin, China; 2007

[7] Zheng J, Fu M. Saturation Control of a Piezoelectric Actuator for Fast Settling-Time Performance. IEEE Transactions on Control Systems Technology. 2013;**21**(1):220-228

[8] Wu Z, Chen M, He P, Li H, Zhang Q, Xiong X, et al. Tracking Control of PZT-Driven CompliantPrecision Positioning Micromanipulator. IEEE Access. 2020;**8**: 126477-126487

[9] Segalman DJ, Starr MJ. Inversion of Masing models via continuous Iwan systems. International Journal of Non-Linear Mechanics. 2008;**43**:74-80

[10] Xiao S, Li Y. Modeling and High Dynamic Compensating the Rate-Dependent Hysteresis of Piezoelectric Actuators via a Novel Modified Inverse Preisach Model. IEEE Transactions on Control Systems Technology. 2012;**21**(5):1549-1557

[11] Lee SH, Royston TJ. Modeling Piezoceramic Transducer Hysteresis in the Structural Vibration Control Problem. The Journal of the Acoustical Society of America. 2000;**108**(6): 2843-2855

[12] Goldfarb M, Celanovic N. Modeling Piezoelectric Stack Actuators for Control of Micromanipulation. IEEE Control Systems Magazine. 1997;**17**(3): 69-79

[13] Liu Y, Shan J, Gabbert U, Qi N. Hysteresis and Creep Modeling and Compensation for a Piezoelectric Actuator Using a Fractional-order Maxwell Resistive Capacitor Approach. Smart Materials and Structures. 2013;**22**: 1-12

[14] PICMA® Stack Multilayer Piezo Actuators [PDF Document]. Physik Instrumente;. Available from: https://www.pi-usa.us/en/products/piezo-actuators-stacks-benders-tubes/p-882-p-888-picma-stack-multilayer-piezo-actuators-100810/#specification.

[15] Shields J, Sirlin S, Wette M. Starlight Pointing Subsystem for the Formation Interferometer Testbed (FIT). Big Sky, MT: IEEE Aerospace Conference; 2002. pp. 1-11

[16] Vishay Precision Group. Transducer-Class Strain Gauges [PDF Document]. Raleigh, NC 27611: Micro-Measurements;. Available from: https: //www.micro-measurements.com.

Chapter 6

Energy Harvesting Prediction from Piezoelectric Materials with a Dynamic System Model

José Carlos de Carvalho Pereira

Abstract

Piezoelectric vibration energy harvesting has been investigated for different applications due to the amount of wasted vibration from dynamic systems. In the case of piezoelectric materials, this energy lost to the environment can be recovered through the vibration of energy harvesting devices, which convert mechanical vibration into useful electrical energy. In this context, this chapter aims to present the mechanical/electrical coupling on a simple dynamic system model in which a linear piezoelectric material model is incorporated. For this purpose, a mechanical/electrical element of a piezoelectric disk is developed and integrated into a lumped mass, viscous damping, and spring assembling, similar to a quarter car suspension system. Equations of motion for this dynamic system in the time domain can be solved using the finite element method. The recovered electric power and energy density for PZT (Lead Zirconate Titanate) from the wasted vibration can be predicted considering that the road roughness is introduced as an input mode.

Keywords: harvesting energy, wasted vibration, dynamic system, linear piezoelectric model, PZT

1. Introduction

Energy use is widely discussed nowadays, as energy conversion and management. In this way, new sources of energy are required to be investigated. Thus, one of the energy sources that can be used is from vibration systems, which can be subject to different excitations. This wasted energy to the environment can be recovered through vibration energy harvesting devices, which convert mechanical vibration into useful electrical energy in a way that low power devices may utilize.

Piezoelectric materials are known to have the electro/mechanical coupling effect. This property has a large range of applications in engineering. Currently, they are extensively used as sensors and actuators in vibration control systems. As a sensor, it can monitor the vibrations when bonded to a flexible structure. As an actuator, it can control the vibration level by introducing a restored force or by adding damping to the system.

In the context of recovered energy from mechanical vibrations based on the conversion of piezoelectric harvesting devices and its application on powering electronic devices, this subject has received the attention from various researchers [1–3]. An example of this type of recovered energy is the suspension system vibration for use of the vehicle itself, such as an energetic source for an active and semi-active suspension.

IntechOpen

On a typical road, vehicles suffer accelerations due to its roughness, which excite undesired vibration. Some recently conducted reviews mentioned the potential of recovering a few hundred watts for a passenger car driven in experimental tests as well as some mathematical models [4, 5]. One of the ways to convert the mechanical energy from the vehicle suspension to electric energy is through piezoelectric materials [6]. Therefore, the objective of this chapter is to present the coupling between a piezoelectric element and a dynamic system in the context of predicting the recovered electric power and energy density for piezoelectric materials, especially the PZT (Lead Zirconate Titanate).

2. Piezoelectric material modelling

Mathematical models for predicting the harvesting energy in piezoelectric materials submitted to axial loads consider its geometric properties, diameter D_p and thickness h_p, and its mechanical and electrical properties, Young's modulus is c_{zz}^E, the piezoelectric constant is e_{zz} and the dielectric constant is ϵ_{zz}^S, as shown in **Figure 1**. Points 1 and 2 represent the two faces of the piezoelectric disk, and w and V are the mechanical displacement and electric potential, respectively, at these two points.

The electromechanical coupling effect of the piezoelectric material can be described using a set of basic equations as given in the IEEE Standard on Piezoelectricity [7]:

$$\begin{aligned} \sigma_z &= c_{zz}^E \frac{\partial w}{\partial z} - e_{zz} \frac{\partial V}{\partial z} \\ D_z &= e_{zz} \frac{\partial w}{\partial z} + \epsilon_{zz}^S \frac{\partial V}{\partial z} \end{aligned} \tag{1}$$

Where σ_z is the normal stress, D_z is the electric flux density, both in direction z. As seen in the above equations, the electro/mechanical coupling occurs due to the piezoelectric constant e_{zz}.

The mechanical strain energy U_m and the electric energy U_e of the piezoelectric material are written as [7]:

$$\begin{aligned} U_m &= \frac{1}{2}\int_V \sigma_z \frac{\partial w}{\partial z} dV = \frac{1}{2}\int_V \left(c_{zz}^E \frac{\partial w}{\partial z} - e_{zz} \frac{\partial V}{\partial z} \right) \frac{\partial w}{\partial z} dV \\ U_e &= \frac{1}{2}\int_V D_z \frac{\partial V}{\partial z} dV = \frac{1}{2}\int_V \left(e_{zz} \frac{\partial w}{\partial z} + \epsilon_z^S \frac{\partial V}{\partial z} \right) \frac{\partial V}{\partial z} dV \end{aligned} \tag{2}$$

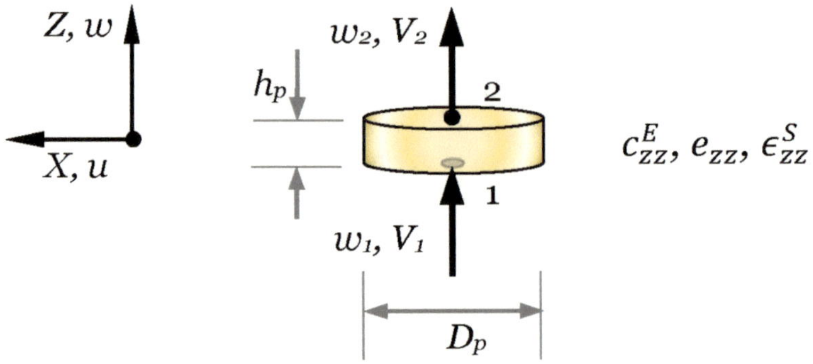

Figure 1.
The piezoelectric material model.

The electrical power can be calculated as the partial derivative of the electric energy, presented by Eq. (2), in respect to time:

$$P_e = \frac{\partial U_e}{\partial t} = \frac{1}{2}\int_V \frac{\partial}{\partial t}\left(D_z \frac{\partial V}{\partial z}\right) dV = \frac{1}{2}\int_V \frac{\partial}{\partial t}\left(e_{zz}\frac{\partial w}{\partial z} + \epsilon_z^S \frac{\partial V}{\partial z}\right)\frac{\partial V}{\partial z} dV \quad (3)$$

3. Dynamic system modelling

This chapter aims to predict the wasted energy from vibration systems that can be further transformed into electrical energy. A typical vibration system can be described as mass, spring and damper elements, and that can represent a suspension system assembly.

The suspension system is an assembly of suspension arms or linkages, springs and shock absorbers that connect the wheels to the vehicle's chassis in order to isolate passengers from vibrations due to bumps and roughness of the road. Furthermore, it must maintain the contact of the wheels with the road to ensure drivability. Thus, the suspension system is the mechanical system where the stability and handling of the vehicle, besides energy harvesting, must be equilibrated.

Mathematical models were initially developed for vertical vehicle performance, and the one-dimensional quarter car model is the simplest from the frequently used suspension system [8]. It is composed of the sprung mass m_s, which represents ¼ of the vehicle's body and the unsprung mass m_u, which represents the wheel assembling mass. Both are considered rigid bodies. Its displacements are w_s and w_u, respectively and both are vertically aligned. There are some studies that include a third degree in the system to describe road roughness excitation w_r. In this case, only the bounce input mode, or the vertical displacement can be implemented. Other elements of the suspension system are included, such as tire stiffness k_t, suspension stiffness k_s, and viscous damping c_s. All vertical displacements are a function of the independent variable t that represents the time. **Figure 2** illustrates this 1D quarter car model, which could represent both the front and rear of the vehicle.

The expressions of kinetical energy from the masses, strain energy and dissipation function from the shock absorber, and the virtual work from the road roughness excitation are written as:

$$\begin{aligned} T &= \frac{1}{2}m_s\dot{w}_s^2 + \frac{1}{2}m_u\dot{w}_u^2 \\ U &= \frac{1}{2}k_s(w_s - w_u)^2 \\ R &= \frac{1}{2}c_s\left(\dot{w}_s^2 - \dot{w}_u^2\right) \\ \delta W &= F_r(t)\delta w_u = k_t w_r(t)\delta w_u \end{aligned} \quad (4)$$

4. The suspension system and piezoelectric disk coupling

The conversion of the mechanical energy from the vehicle suspension to electric energy through piezoelectric materials can be predicted by piezoelectric disk coupling illustrated in **Figure 1** and the suspension system illustrated in **Figure 2**. Since the conversion of mechanical energy to electrical energy in this system is produced by compressive efforts, the piezoelectric disk is located between the shock absorber

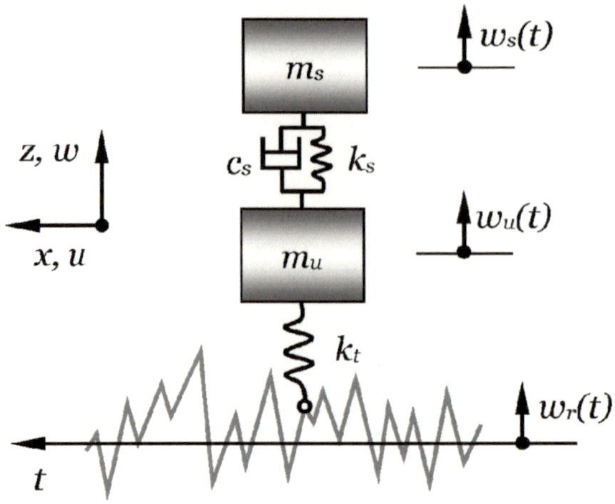

Figure 2.
1D quarter car model.

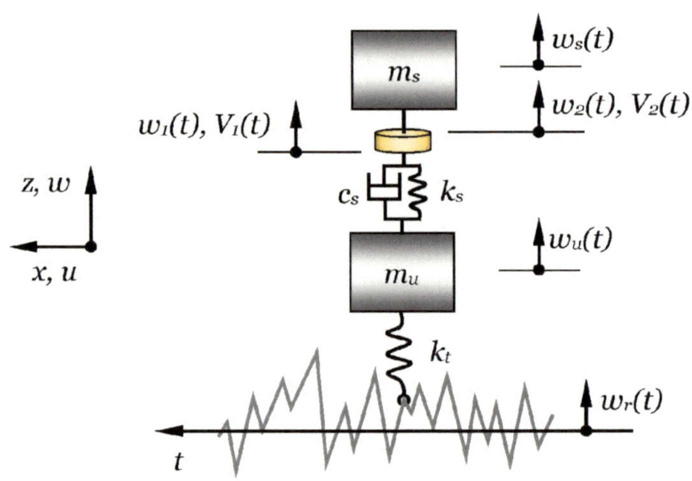

Figure 3.
The suspension system and piezoelectric coupling model.

system composed by stiffness k_s and viscous damping c_s on the bottom side, along with the sprung mass m_s on the upper side, as illustrated in **Figure 3**. As stated previously, vertical displacements w and now the electric potential V, are all a function of the independent variable t.

5. The vertical displacement and electric potential approach

Understanding physical problems can be accomplished when the numerical simulation of equations within the variables that describe them are represented. In the case of the electric energy prediction from the vehicle suspension system, the variables are vertical displacement w and electric potential V of different points, or nodes, and the domain within the model is considered valid.

Some numerical methods are often used to understand physical problems, and among them, the most widely used in engineering is the finite element method. Within this method, the variables in the equations that describe the physical problem must be approximated to the polynomial functions of these variables. For the application of this method in solving problems, all elements that compose it must be represented by matrices, called elementary matrices, which are the result of the adopted polynomial functions. For further details on this method, the following references [9, 10] are recommended.

If the linear approximation function of vertical displacement w and electric potential V over thickness h_p in direction z of the piezoelectric disk are considered, stiffness elementary matrices of the piezoelectric finite element can be obtained by applying the Lagrange equations [11] over the energy expressions presented by Eq. (2).

$$
\begin{aligned}
[K_m] &= \frac{c_{zz}^E S_p}{h_p}\begin{bmatrix} 1 & -1 \\ -1 & 1 \end{bmatrix} = k_m \begin{bmatrix} 1 & -1 \\ -1 & 1 \end{bmatrix} \\
[K_{m\_el}] &= -\frac{e_{zz} S_p}{2h_p}\begin{bmatrix} 1 & -1 \\ -1 & 1 \end{bmatrix} = -k_{m\_el} \begin{bmatrix} 1 & -1 \\ -1 & 1 \end{bmatrix} \\
[K_{el}] &= \frac{\epsilon_z^S S_p}{h_p}\begin{bmatrix} 1 & -1 \\ -1 & 1 \end{bmatrix} = k_{el} \begin{bmatrix} 1 & -1 \\ -1 & 1 \end{bmatrix}
\end{aligned} \tag{5}
$$

Where $[K_m]$, $[K_{m\_el}]$ and $[K_{el}]$ are mechanical elementary stiffness, electromechanical coupling elementary stiffness and electric elementary stiffness matrices, respectively, and S_p is the cross-sectional area of the piezoelectric disk.

Moreover, if the same linear approximation function of the vertical displacement w over stiffness k_s and viscous damping c_s of the shock absorber in direction z are considered, the suspension system's differential equation of motion, illustrated in **Figure 2**, can be obtained by applying the Lagrange equations [11] over the energy expressions presented by Eq. (4).

$$
[M]\{\ddot{w}\} + [C]\{\dot{w}\} + [K]\{w\} = \{F(t)\} \tag{6}
$$

Where the elementary matrices and the vectors are:

$$
\begin{aligned}
[M] &= \begin{bmatrix} m_s & 0 \\ 0 & m_u \end{bmatrix} \\
[K] &= \begin{bmatrix} k_s & -k_s \\ -k_s & k_s \end{bmatrix} \\
[C] &= \begin{bmatrix} c_s & 0 \\ 0 & c_s \end{bmatrix} \\
\{w\} &= \begin{Bmatrix} w_s(t) \\ w_u(t) \end{Bmatrix} \\
\{\dot{w}\} &= \begin{Bmatrix} \dot{w}_s(t) \\ \dot{w}_u(t) \end{Bmatrix} \\
\{\ddot{w}\} &= \begin{Bmatrix} \ddot{w}_s(t) \\ \ddot{w}_u(t) \end{Bmatrix} \\
\{F(t)\} &= \begin{Bmatrix} 0 \\ k_t w_r(t) \end{Bmatrix}
\end{aligned} \tag{7}
$$

Using the technique of assembling the elementary matrices of the finite element method, the suspension system's differential equation of motion and piezoelectric disk coupled model, as illustrated in **Figure 3**, is as shown in Eq. (8):

$$[M_m]\{\ddot{w}\} + [C_m]\{\dot{w}\} + [K_m]\{w\} + [K_{m\_el}]\{V\} = \{F(t)\} \quad (8)$$

$$[K_{m\_el}]^t\{w\} + [K_{el}]\{V\} = \{0\} \quad (9)$$

Eq. (9) can be manipulated and substituted for Eq. (8). Thus, the final equation of motion as a function of only mechanical variables is:

$$[M_m]\{\ddot{w}\} + [C_m]\{\dot{w}\} + \left[[K_m] - [K_{m\_el}][K_{el}]^{-1}[K_{m\_el}]^t\right]\{w\} = \{F(t)\} \quad (10)$$

Eq. (10) can be solved in the time domain with an integration method, such as the Newmark Method [12]. The mechanical force is due to road roughness as well as the tire characteristics of the wheel as shown in Eq. (7). These data are used to apply the condition in each time step in solving Eq. (10), in which all variables $w_s = w_2$, w_1, and w_u and their time derivatives are obtained.

The response in the time domain in respect to vertical displacements is obtained by using Eq. (10) and subsequently, the electric potential is obtained as:

$$\{V(t)\} = -[K_{el}]^{-1}[K_{mel}]^t\{w(t)\} \quad (11)$$

The electrical energy can be calculated as presented by Eq. (2). The development of this equation follows:

$$U_e = \frac{1}{2}\int_0^{h_p}\int_{S_p}\left[e_{zz}\left(\frac{\partial w}{\partial z}\right)\frac{\partial V}{\partial z} + \epsilon_z^S\left(\frac{\partial V}{\partial z}\right)^2\right]dS\ dz \quad (12)$$

Thus, the expression of the electrical energy due to the piezoelectric disk on this suspension system is as:

$$U_e = \frac{1}{2}\frac{e_{zz}S_p}{h_p}\ (w_2 - w_1)\ (V_2 - V_1) + \frac{\epsilon_z^S S_p}{h_p}\ (V_2 - V_1)^2 \quad (13)$$

In addition, the electrical power can be calculated as presented by Eq. (3) and its development is as follows:

$$P_e = \frac{1}{2}\int_0^{h_p}\int_{S_p}\left[e_{zz}\ \left(\frac{\partial}{\partial t}\left(\frac{\partial w}{\partial z}\right)\ \frac{\partial V}{\partial z} + \frac{\partial w}{\partial z}\ \frac{\partial}{\partial t}\left(\frac{\partial V}{\partial z}\right)\right) + \epsilon_z^S\ \frac{\partial}{\partial t}\left(\frac{\partial V}{\partial z}\right)^2\right]dS\ dz \quad (14)$$

Thus, the expression of the electrical power due to the piezoelectric disk on this suspension system is as:

$$P_e = \frac{1}{2}\frac{e_{zz}S_p}{h_p}\ \left[(\dot{w}_2 - \dot{w}_1)\ (V_2 - V_1) + (w_2 - w_1)\ (\dot{V}_2 - \dot{V}_1)\right] + \frac{\epsilon_z^S S_p}{h_p}\ (V_2 - V_1)\ (\dot{V}_2 - \dot{V}_1) \quad (15)$$

Where $\dot{w}_1$, $\dot{w}_2$, $\dot{V}_1$ and $\dot{V}_2$ are time derivatives of the displacements and electric potential of nodes 1 and 2 of the piezoelectric disk.

6. Application example

For a simpler demonstration of the potential for harvesting energy in a suspension system by means of piezoelectric material, a MATLAB® code to obtain the results was developed and presented below.

The dimensions of the piezoelectric disk are diameter $D_p = 0.065\,m$ and thickness $h_p = 0.025\,m$. The vehicle data and the properties of piezoelectric material PZT-5H are shown in **Tables 1** and **2**, as indicated in references [13, 14], respectively.

Property description	Value
Body and wheel mass	
¼ Vehicle sprung mass (m_u)	362.5 kg
Unsprung mass (m_s)	39 kg
Tire stiffness (k_t)	200 kN/m
Shock absorber suspension	
Suspension stiffness (k_s)	30 kN/m
Suspension damping (c_s)	4 kN s/m

Table 1.
Vehicle data.

	PZT-5H
Piezoelectric constant – e_{zz} [C/m^2]	23.30
Dielectric constant – ϵ_{zz}^{S} [F/m]	1.30 x 10^{-8}
Density – ρ [kg/m^3]	7500
Young's modulus – c_{zz}^{E} [GPa]	23.0

Table 2.
Properties of piezoelectric material.

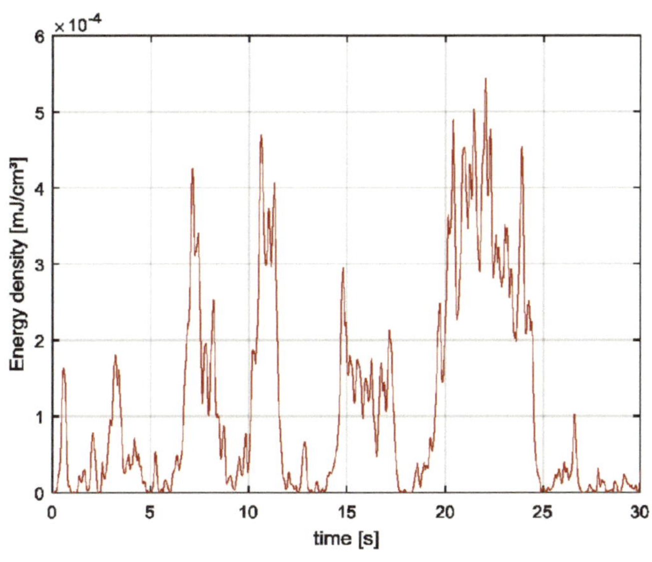

Figure 4.
Electric energy density response for PZT-5H.

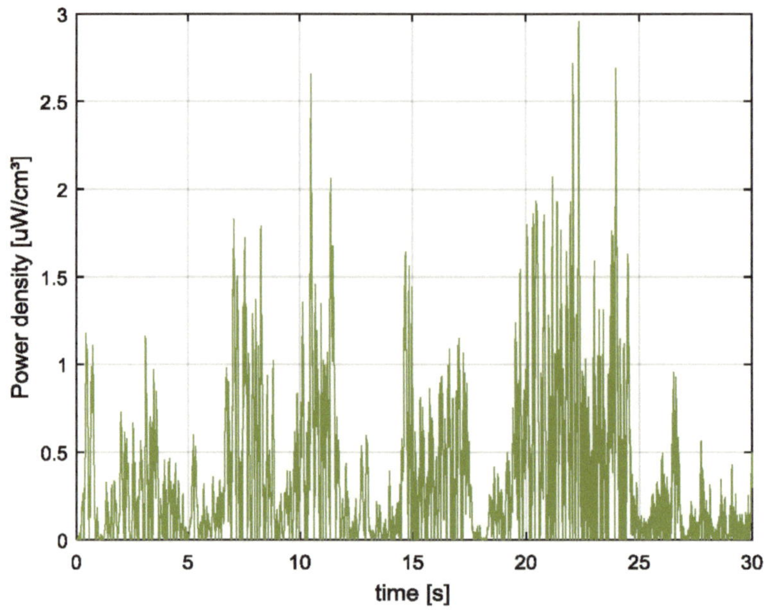

Figure 5.
Electric power density response for PZT-5H.

The road excitation is as depicted in [6]. The excitation is the bounce input mode vertical position of the road $w_r(t)$ with class D road (poor) and a driving speed for a vehicle of 20 m/s.

In **Figure 4**, the instantaneous electric energy density for piezoelectric material PZT-5H calculated by Eq. (13), with a simulation time of 30 s and a fixed step time of 0.00001 s.

Figure 5 exhibits the overall instantaneous electric power density response for piezoelectric material PZT-5H calculated by Eq. (15).

7. Conclusions

This chapter proposes a coupled suspension system and piezoelectric model to predict the potential of harvested electric power in vehicle suspension systems. The performance of piezoelectric material PZT-5H was investigated, in respect to harvesting energy based on energy density and electric power density.

The approach presented in this chapter is a way to simulate the electric power generated in vehicle suspension systems established by piezoelectric harvesting. Nonetheless, these results would need to be compared with experimental results to demonstrate the validity of the proposed model.

Author details

José Carlos de Carvalho Pereira
Mechanical Engineering Department, Federal University of Santa Catarina, Florianópolis, Brazil

*Address all correspondence to: carlosp@emc.ufsc.br

IntechOpen

© 2021 The Author(s). Licensee IntechOpen. This chapter is distributed under the terms of the Creative Commons Attribution License (http://creativecommons.org/licenses/by/3.0), which permits unrestricted use, distribution, and reproduction in any medium, provided the original work is properly cited.

References

[1] Shenck NS, Paradiso JA. Energy scavenging with shoe-mounted piezoelectrics. IEEE Micro. 2001;**21**: 30-41

[2] Ottman GK, Hofmann HF, Lesieutre GA. Optimized piezoelectric energy harvesting circuit using step-down converter in discontinuous conduction mode. IEEE Transactions Power Electronics. 2003;**18**:696-703

[3] Roundy S, Wright PK. A piezoelectric vibration based generator for wireless electronics. Smart Materials and Structures. 2004;**13**:1131-1142

[4] Zhang Y, Guo K, Wang D, et al. Energy conversion mechanism and regenerative potential of vehicle suspensions. Energy. 2017;**119**:961-970

[5] Addelkareem MAA, Xu L, Guo X, et al. Energy harvesting sensitivity analysis and assessment of the potential power and full car dynamics for different road modes. Mechanical Systems and Signal. 2018;**110**:307-332

[6] Morangueira YLA, Pereira JCC. Energy Harvesting Assessment with a Coupled Full Car and Piezoelectric Model. Energy. 2020;**210**:1-13

[7] IEEE Standard on Piezoelectricity. ANSI Standard 176; 1987.

[8] Zuo L, Zhang P. Energy harvesting, ride comfort, and road handling of regenerative vehicle suspensions. Journal of Vibration and Acoustics. 2013; 135: 011002–1–011002-8.

[9] Yang TY. Finite Element Structural Analysis. ed. In: Prentice-Hall. 1986

[10] Bathe KJ. Finite Element Procedure. ed. In: Prentice Hall. 1996

[11] Meirovitch L. Methods of analytical Dynamics. ed. McGraw-Hill, Inc.; 1988.

[12] Pereira JCC. Fundamentos da Análise de Sistemas Mecânicos. ed. In: UFSC. 2017

[13] Addelkareem MAA, Xu L, Guo X, et al. Energy Harvesting Sensitivity Analysis and Assessment of the Potential Power and Full Car Dynamics For Different Road Modes. Mechanical Systems and Signal. 2018;**110**:307-332

[14] Kocbach J. Finite Element Modeling of Ultrasonic Piezoelectric Transducers. University of Bergen. Tech. Rep; 2000

Section 4

Applications

Chapter 7

Design, Characterisation and Prospect of Piezoelectric Microfluidic Technology

Song Chen, Zhonghua Zhang, Junwu Kan, Jianping Li and Jianming Wen

Abstract

Fluidic driving device plays an important role in the delivery and distribution of minute amount of the liquid in the micro-fluidic system. Due to the unique advantages of simple structure, short response time, and low power consumption, piezoelectric actuation was employed to implement the microfluidic transportation. A piezoelectrically driven microfluidic device, piezoelectric pump, was developed and widely applied in many fields in last three decades. As a kind of displacement pump, piezoelectric pumps is able to realize accurate transportation of the liquid because of per stroke of output fluid is equal to the volumetric change of pumping chamber. And the output flow rate and pressure is easily to be controlled through adjusting the driving voltage or frequency. In this chapter, the design, structure, working principle and the characterisation of piezoelectric pumps with single chamber and multiple chambers are introduced.

Keywords: piezoelectric actuator, microfluidic technology, micropump, single chamber, multiple chamber, flow rate, pressure

1. Introduction

Microfluidic modules have broad application prospects in chemical analysis, biomaterial analysis, fuel cell, and medical [1, 2]. In the microfluidic system, the fluid pump is the core to realize small and accurate liquid delivery and distribution [3, 4]. Therefore, a variety kinds of micropumps have been developed [5–8]. Because of its simple structure, high energy density and simple control, piezoelectric pump has potential application prospects in drug delivery, chemical analysis, micro refrigeration system and micro spacecraft propulsion [9–11].

According to the driving components, piezoelectric pump can be divided into piezoelectric stack pump and piezoelectric diaphragm pump. The piezoelectric diaphragm pump owns superiorities of small volume and low power consumption. This chapter will take the piezoelectric diaphragm pump as the example to introduce the piezoelectric microfluidic technology.

IntechOpen

2. Piezoelectric pump with single chamber

The piezoelectric pump with single chamber is the basic form of piezoelectric pump. The one-way fluid drive of piezoelectric pump is realized by matching with check valves. The common piezoelectric pumps with single chamber are that with single wafer, bimorph and double actuator.

2.1 Piezoelectric pump with single chamber

Figure 1 shows the structure and working principle of the piezoelectric pump with single chamber and single wafer (PPSCSW). Single wafer piezoelectric vibrator, pump body and two check valves (inlet and outlet) constitute a closed pump chamber. The volume of the pump chamber changes with the reciprocating motion of piezoelectric diaphragm driven by AC voltage signal. The inlet valve and outlet valve are opened and closed periodically to form suction and discharge. At the suction stage, the volume of the pump chamber increases, the inlet valve opens, the outlet valve closes, and the fluid enters the pump chamber. When discharging, the volume of the pump chamber is reduced, the outlet valve is opened, the inlet valve is closed, and the fluid is discharged from the pump chamber. This circulation enables one-way flow of fluid.

Figures 2 and **3** show the three-dimensional structure and prototype of PPSCSW. The shell and pump body are made of the polymethylmethacrylate (PMMA), which has superiorities of easy processing and transparent material. The dimension of upper, middle and bottom shells is 40 mm × 40 mm × 5.5 mm. The diameter of outlet pipe and inlet pipe is Φ7 mm. The inlet and outlet check valves made of rubber umbrella are installed in the same place, with the size of Φ10 mm × 0.5 mm. By adjusting the overall size to 40 mm × 40 mm × 16.5 mm, PPSCSW can precisely control the fluid output flow rate and pressure.

Figure 4 is the relationships between output flowrate of PPSCSW and the driving voltage and frequency. When the driving voltage is fixed at 150 V_{pp} and 75 V_{pp}, the maximum flow rates of 5.42 ml/min and 26.55 ml/min are obtained at 20 Hz and 30 Hz, respectively. When the driving frequency is fixed at 15 Hz and 30 Hz, the flow rate increases with the increase of the driving voltage. When the driving voltage increases to 210 V_{pp}, the output flow rate of PPSCSW reaches 26.55 ml/min.

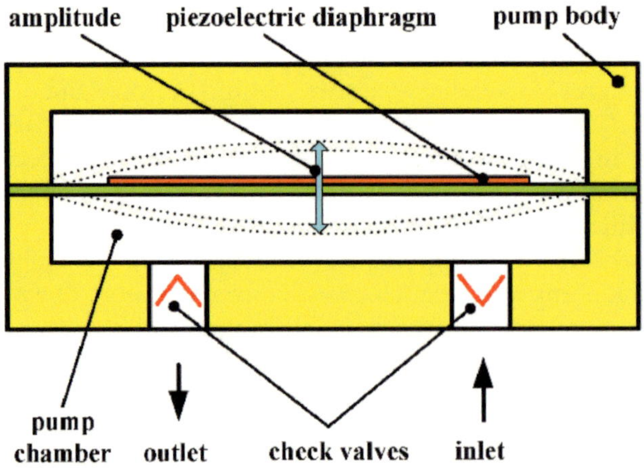

Figure 1.
The structure and working principle of PPSCSW [12].

Figure 2.
The three-dimensional structure of PPSCSW [12].

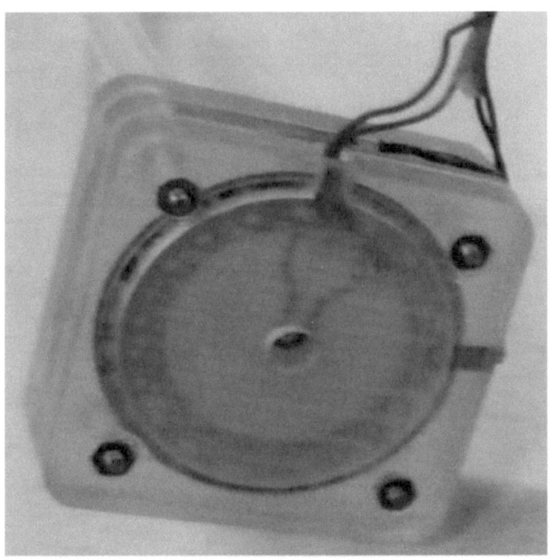

Figure 3.
The prototype of PPSCSW [12].

2.2 Piezoelectric pump with single chamber and a bimorph actuator

As the name implies, the bimorph actuator is composed of two PZT disks with mechanical serial and parallel structure. **Figure 5** shows the structure of a piezoelectric pump with single chamber and a bimorph actuator (PPSCBA). The working principle of PPSCB is similar to that of PPSCSW in **Figure 1**. However, compared with single wafer piezoelectric pump, double piezoelectric actuators usually have symmetrical structures. In theory, PPSCBA can monitor the output performance by using a PZT disk as a monitor, which will make it more widely application.

In order to study the PPSCBA, the three-dimensional structure and prototype of PPSCBA have been designed, as shown in **Figures 6** and **7**. A bimorph is used as the actuator and sensor. The check valve is an umbrella valve made of rubber.

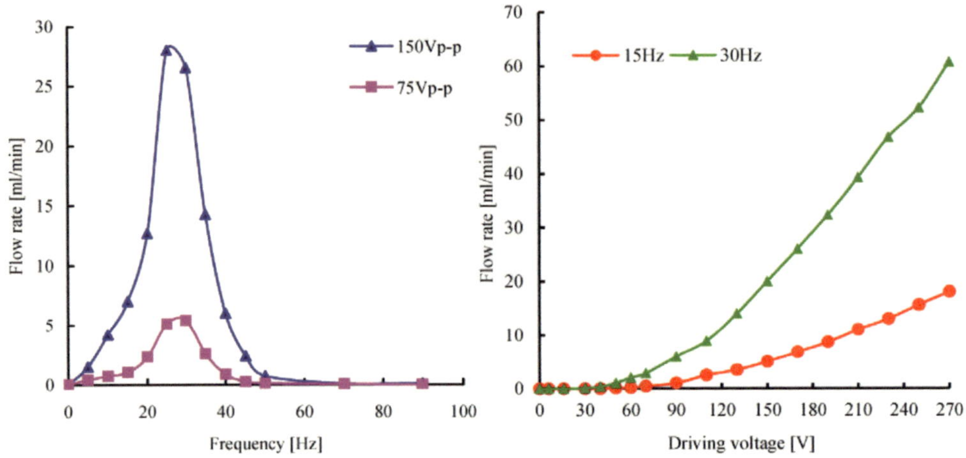

Figure 4.
The relationship between flowrate and driving frequency and voltage [12].

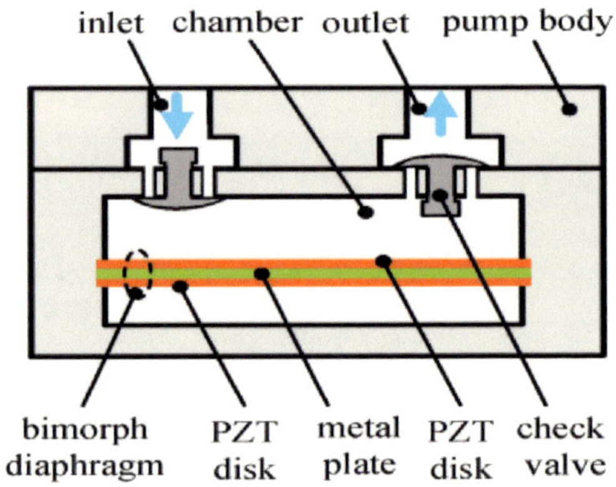

Figure 5.
The structure of PPSCBA [13].

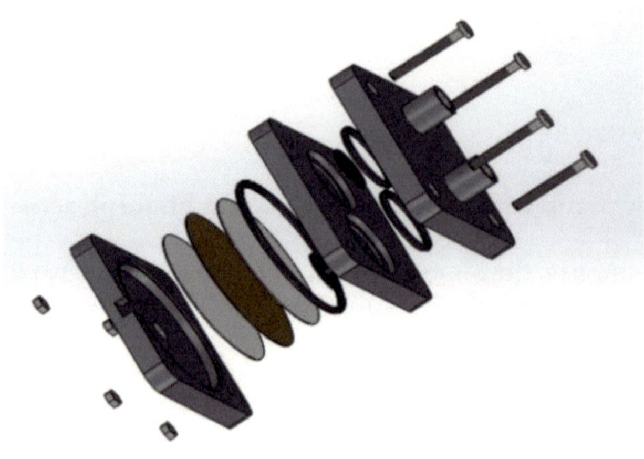

Figure 6.
The three-dimensional structure of PPSCBA [13].

Figure 7.
The prototype of PPSCBA [13].

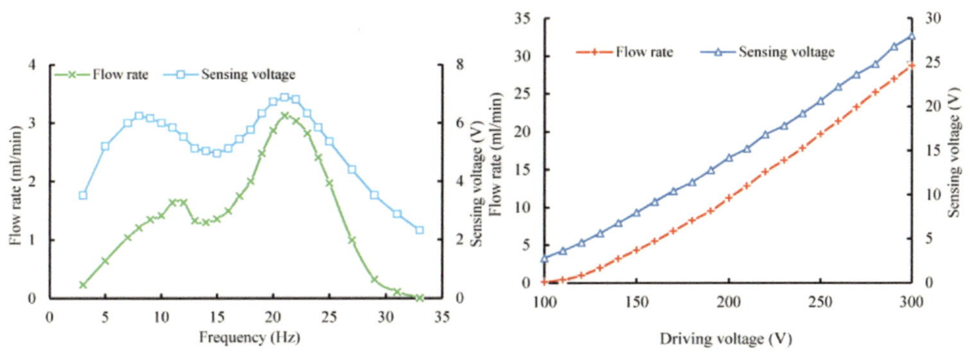

Figure 8.
The relationship between flowrate and driving frequency and voltage [13].

The material of pump body is PMMA. The overall dimension of PPSCBA is 40 mm × 40 mm × 17 mm. When one PZT disc is used as the driver and the other as the monitor, the monitor of output flow rate is achieved.

As shown in **Figure 8**, when one piezoelectric wafer of the bimorph piezoelectric vibrator is used as the driver and the other as the sensor, the relationship between the output flow rate and the sensing voltage of PPSCBA and the driving frequency and voltage is analyzed. When the driving voltage is fixed at 150 V_{pp}, the peak values of sensing voltage and output flow, 6.88 V_{pp} and 3.12 ml/min, are obtained at 21 Hz. Therefore, the optimal frequency of PPSCBA is able to realize self-sensing. When the driving frequency is fixed at 21 Hz, the flow rate and sensing voltage increase with the increase of driving voltage, and the increasing trend is similar. When the driving voltage increases to 300 V_{pp}, the output flow rate and sensing voltage reach the maximum, which are 28.71 ml/min and 28.0 V_{pp} respectively. Therefore, PPSCBA can realize the self-sensing of output flow rate.

2.3 Piezoelectric pump with single chamber and double piezoelectric actuator

In order to improve the drive ability of pump chamber, the piezoelectric pump with single chamber and double piezoelectric actuator (PPSCDPA) is proposed. **Figure 9** shows the structure of PPSCDPA. PPSCDPA consists of an inlet, two

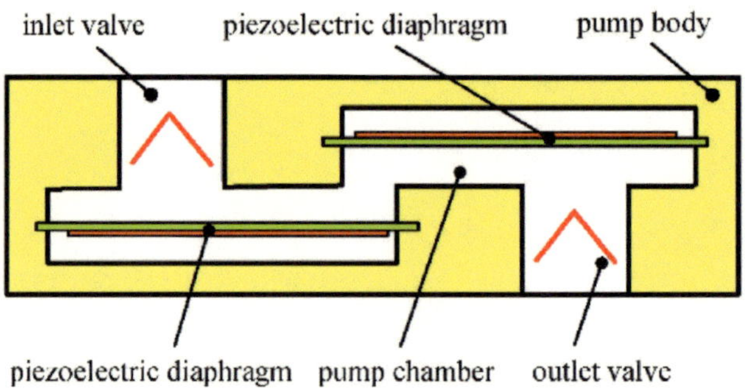

Figure 9.
The structure of PPSCDPA [14].

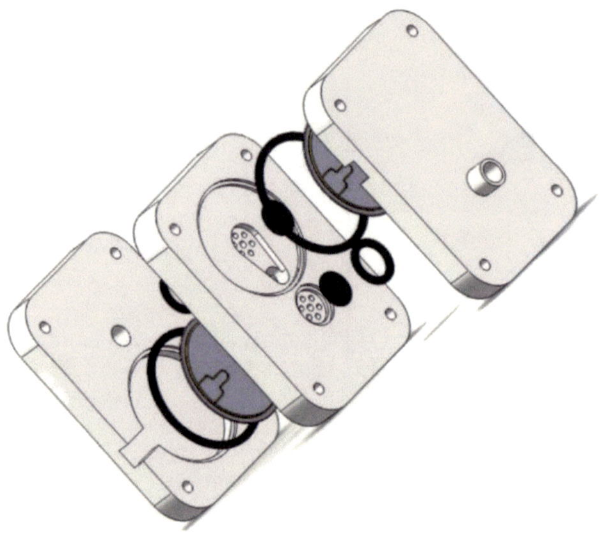

Figure 10.
The 3-D structure of PPSCDPA [14].

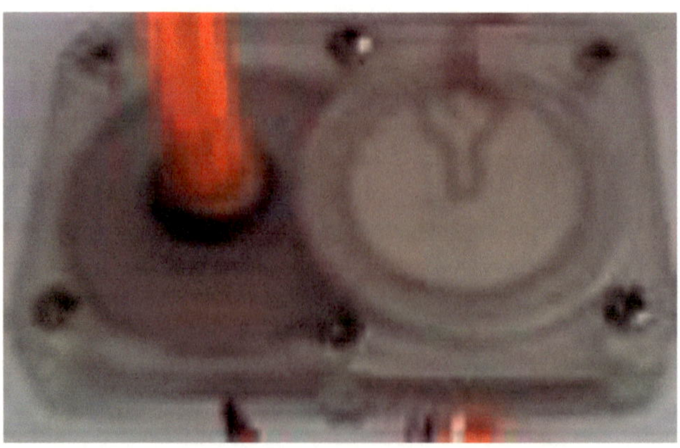

Figure 11.
The prototype of PPSCDPA [14].

piezoelectric diaphragm, two check valves, a pump chamber, a pump body and an outlet. The vibration of the two piezoelectric diaphragms is the same, that is, their driving voltage signals are the same. In the process of suction, two piezoelectric diaphragms bend out of the chamber at the same time; while discharging, the two piezoelectric diaphragms bend to the chamber at the same time. Combined with the unidirectional characteristic of the check valve, the PPSCDPA can realize the one-way transportation of fluid from the inlet to the outlet under the AC voltage signal drive.

The three-dimensional structure and prototype of PPSCDPA with the overall dimension of 65 mm × 40 mm × 12 mm are shown in **Figures 10** and **11**, respectively. The check valves are umbrella valve made of rubber. The material of pump body is PMMA. The performance test platform of PPSCDPA is established, and the output performance of piezoelectric pump is tested. The experimental results show that the maximum flow rate of 45.98 ml/min is achieved when the driving parameters are 15 Hz and 200 V_{pp}.

3. Piezoelectric pump with multiple chambers

The output performance of the piezoelectric pump can be improved by increasing the number of chambers. When the pump chambers are connected in series, the pressure of the piezoelectric pump increases; when the pump chambers are parallel, the flow rate of the piezoelectric pump increases.

3.1 Piezoelectric pump with single actuator and double chamber

As shown in **Figure 12**, the piezoelectric pump with single actuator and double chamber (PPSADC) includes a piezoelectric bimorph vibrator, two chambers (chamber A and B), four check valves and a pump body. Chamber A has inlet 1 and outlet 1, and chamber B has inlet 2 and outlet 2. The chamber A and chamber B of the piezoelectric pump can be connected in series or in parallel, which can improve the applicable range of the pressure and flow. **Figure 12(a)** shows the series connection of chambers A and B. In the process of working, the fluid enters the chamber B from the inlet 2, and the outlet 2 is connected with the inlet 1, and the liquid enters the chamber A through the inlet 1, so that chamber B and chamber A are in series, which can make the piezoelectric pump obtain high pressure. **Figure 12(b)** shows the parallel connection of chamber A and B. In the process of working, the fluid enters the chamber A and B from the inlet 1 and 2, and flows out from the outlet 1 and 2, so that the chamber A and B form a parallel connection, which can obtain a large flow rate.

Figure 13 is the three-dimensional structure of PPSADC with the overall dimension of 41 mm × 41 mm × 25 mm. **Figure 14** is the prototype of PPSADC. Chamber A and chamber B are symmetrical. The upper and lower O-rings are used to seal the chambers, which realizes the flexible support of the piezoelectric vibrator. The flexible support can amplify the displacement of the piezoelectric vibrator. The structure of check valve is umbrella valve. The material of pump body is PMMA. The series and parallel switching of chamber A and chamber B is realized through the connection transformation of rubber tube.

Figure 15 shows the relationship between the output flow rate and the driving voltage when the driving frequency is 100 Hz. It can be seen from the **Figure 15** that the output flow increases with the increase of the driving voltage. When the driving

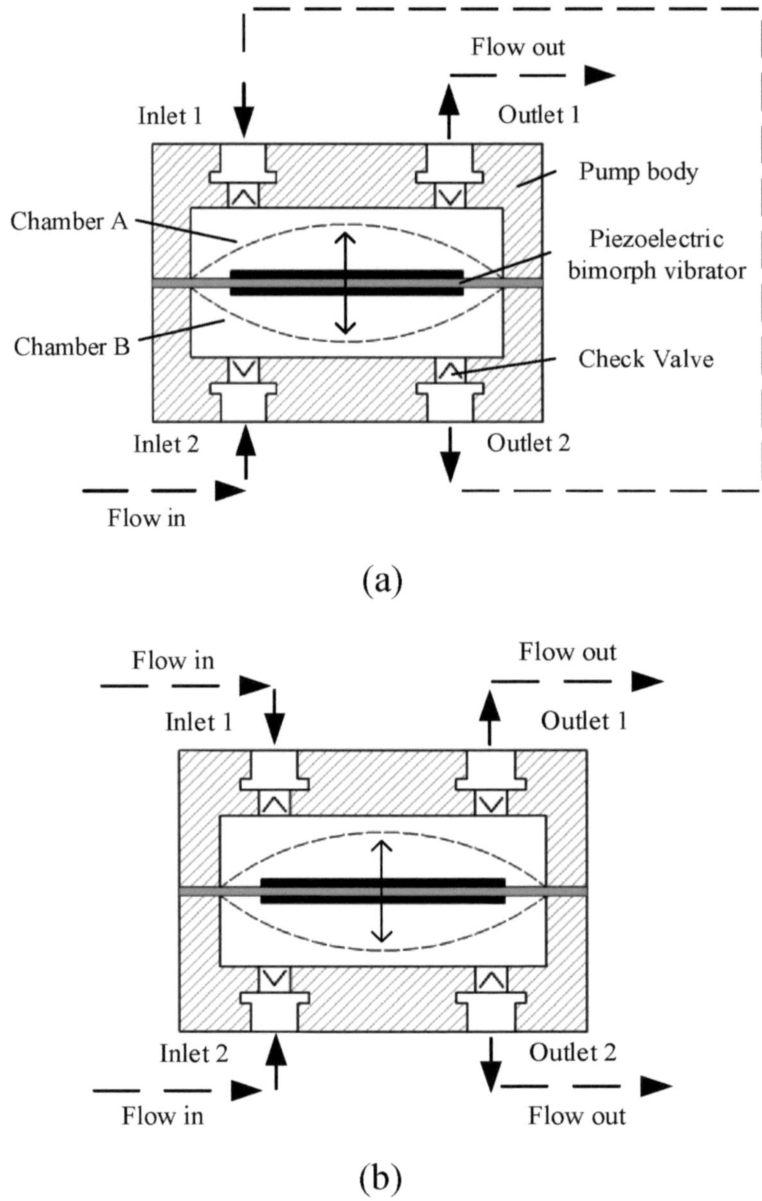

Figure 12.
The structure and working principle of PPSADC. (a) Serial connection. (b) Parallel connection [15].

voltage is increased to 250 V_{pp}, the output flow rate of the two chambers in series and in parallel is 17 ml/min and 32 ml/min respectively. The PPSADC can achieve high energy conversion efficiency.

3.2 Piezoelectric pump with five actuators and five chambers

Figure 16 is the structure of piezoelectric pump with five actuators and five chambers (PPFAFC). Five piezoelectric actuators and five chambers constitute five pumps in series, which improves the output pressure. A_i (i = 1, 2, 3, 4, 5) represents five piezoelectric actuators; C_j (j = 1, 2, 3, 4, 5) represents five chambers; V_k (k = 1, 2, 3, 4, 5, 6) represents 6 check valves (**Figure 16**).

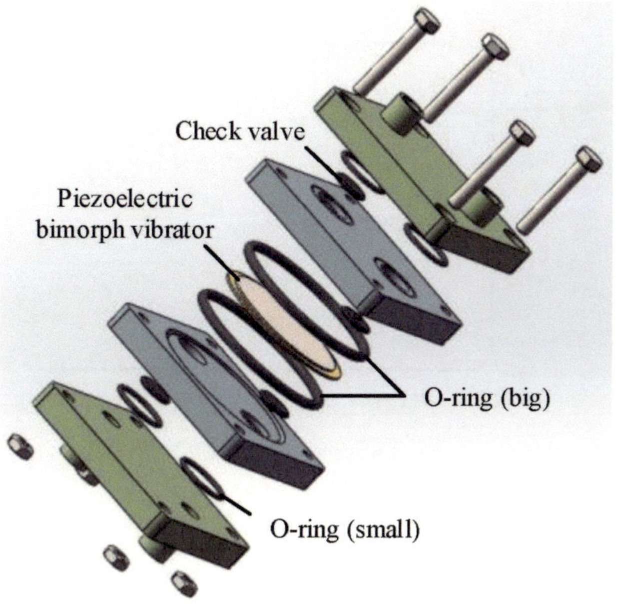

Figure 13.
The 3-D structure of PPSADC [15].

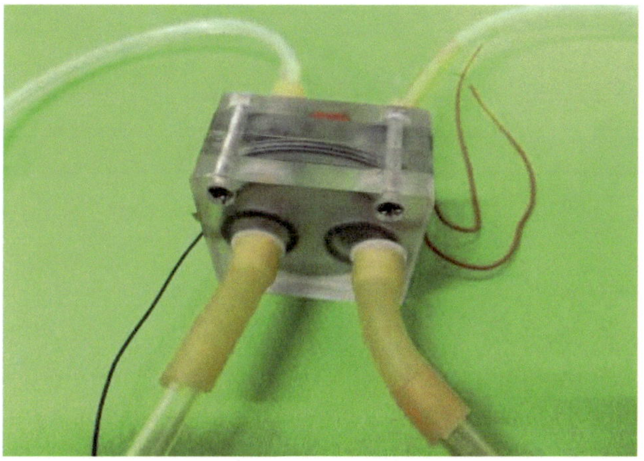

Figure 14.
The prototype of PPSADC [15].

Figure 17 shows the working principle of PPFAFC. PPFAFC is driven by an AC voltage signal, and the driving voltage of the adjacent piezoelectric vibrator has a phase angle of 180°. As shown in **Figure 14(a)**, during the suction process, A1/A3/A5 bends outwards of the cavity and A2/A4 bends into the cavity. Then the volume of the chamber C1/C3/C5 increases, and the volume of the chamber C2/C4 decrease, valve V1/V3/V5 opens, valve V2/V4 closes; as shown in **Figure 14(b)**, during the discharge process, A2/A4 bends outwards of the chamber and A1/A3/A5 bends towards the chamber, then the volume of the C2/C4 increases, the volume of C1/C3/C5 decreases, the valve V2/V4 opens, and the valve V1/V3/V5 closes. When it is driven by an AC voltage signal, the suction and discharge processes are alternately performed to achieve continuous fluid output (**Figure 17**).

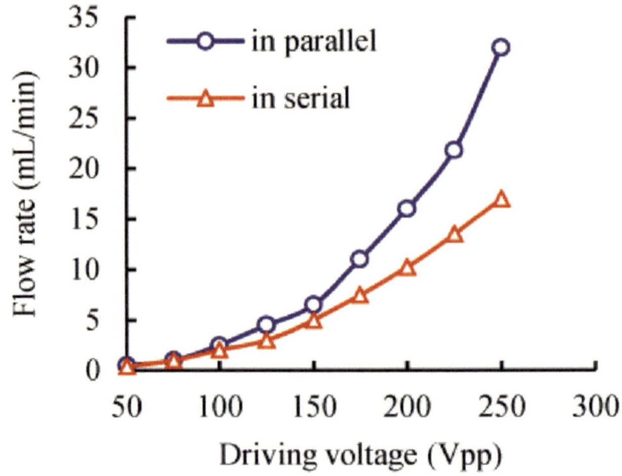

Figure 15.
The relationship between flow rate and driving voltage [15].

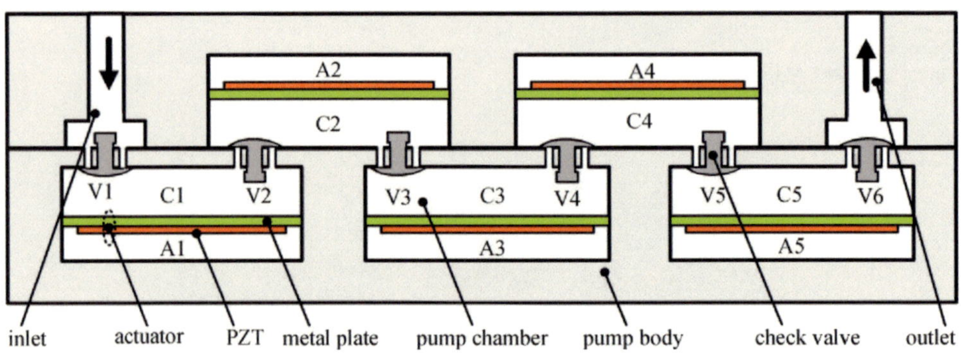

Figure 16.
The structure of PPFAFC [16].

Figure 15 shows the three-dimensional structure and prototype of PPFAFC. A prototype of PPFAFC was designed, manufactured and assembled. The performance of PPFAFC is tested. The pump body is made of PMMA and sealed by O-ring. It was driven by five piezoelectric actuators. PPFAFC can achieve high output performance at low voltage, where it can meet the needs of different applications (**Figure 18**).

As shown in **Figure 19**, when the driving voltage is 180 V_{pp}, the relationship between the output flow rate and backpressure of PPFAFC and the driving frequency is analyzed. It can be seen from **Figure 19** that the output flow rate increases with the increase of the driving frequency. When the driving frequency increases to 400 Hz, the flow rate increases to 279.2 ml/min. When the driving frequency is 90 Hz, the maximum output pressure of 10.8 kPa is obtained. PPFAFC owns high output performance.

3.3 Piezoelectric pump with five actuators and ten chambers

Figures 20 and **21** show the three-dimensional structure and working principle of piezoelectric pump with five actuators and ten chambers (PPFATC). PPFATC is composed of five groups of single vibrator double chamber piezoelectric pumps in

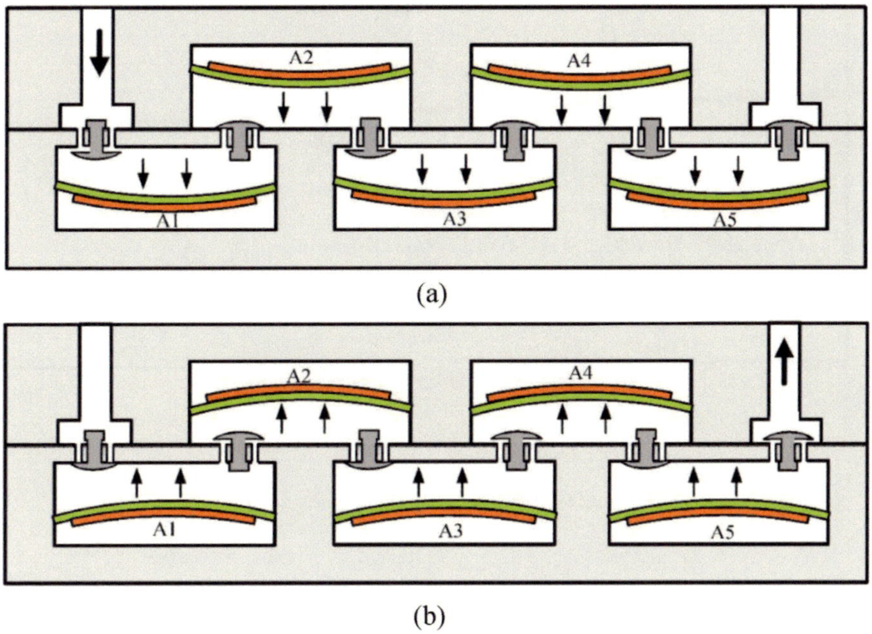

Figure 17.
Working principle of PPFAFC. (a) Suction. (b) Discharge [16].

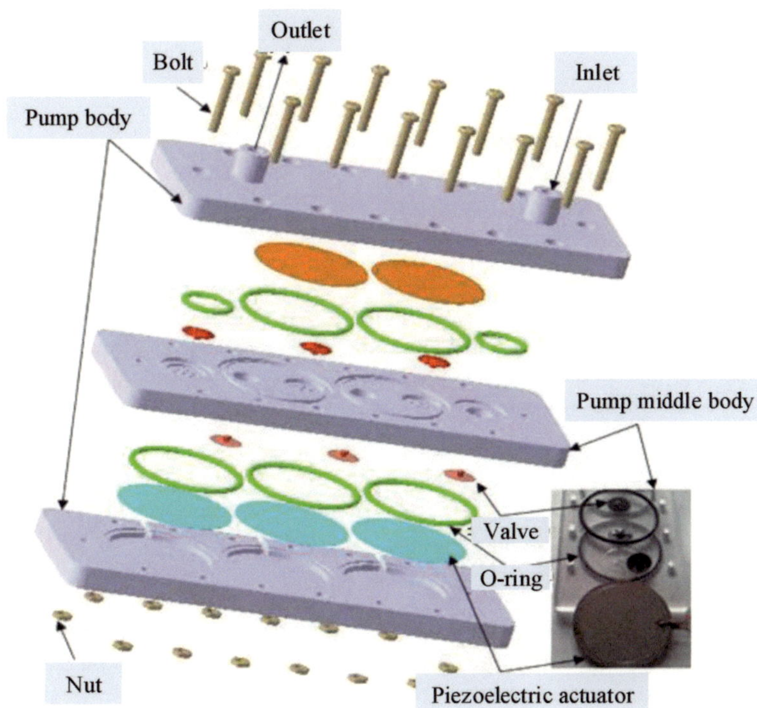

Figure 18.
The three-dimensional structure and prototype of PPFAFC [16].

series, forming two group of five-chambers mechanisms. When PPFATC is connected in parallel, two groups of five-chambers mechanisms are formed in parallel, which can make the outlet obtain continuous fluid output without fluctuation;

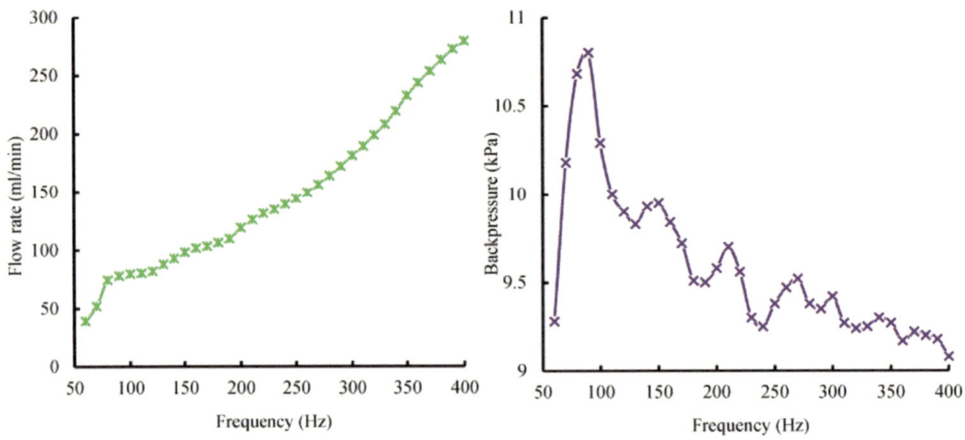

Figure 19.
The relationship between flow rate and backpressure and excitation frequency [16].

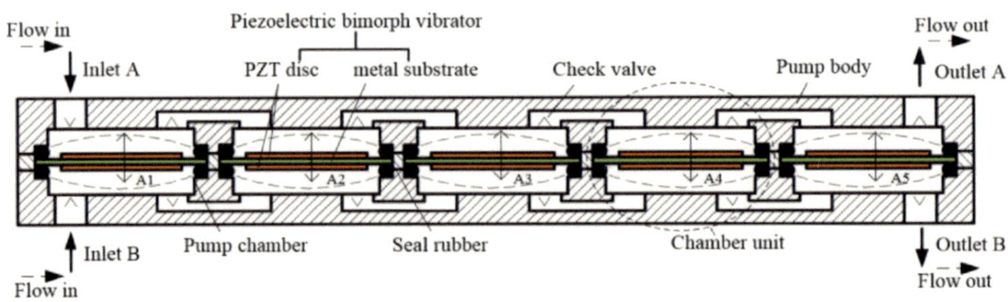

Figure 20.
The structure and working principle of PPFATC in parallel connection [17].

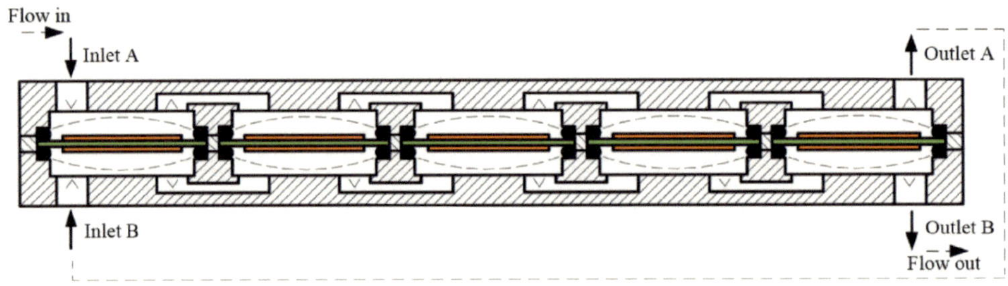

Figure 21.
The structure and working principle of PPFATC in serial connection [17].

when PPFATC is connected in series, two groups of five-chambers mechanisms are formed in series, namely ten chambers in series, which can make the output pressure increase. In the working process, the phase of the driving voltage signal of the adjacent piezoelectric vibrator is 180 degrees, that is, the vibration direction of the adjacent piezoelectric vibrator is opposite in the working process. Combined with the function of check valve, the continuous one-way fluid output of PPFATC is formed under the driving of AC voltage signal.

Figure 22 is the three-dimensional structure and prototype of the PPFATC. A prototype of the PPFATC is fabricated, and then carried out its output performance test and water cooling test with the PPFATC as the power source, as shown

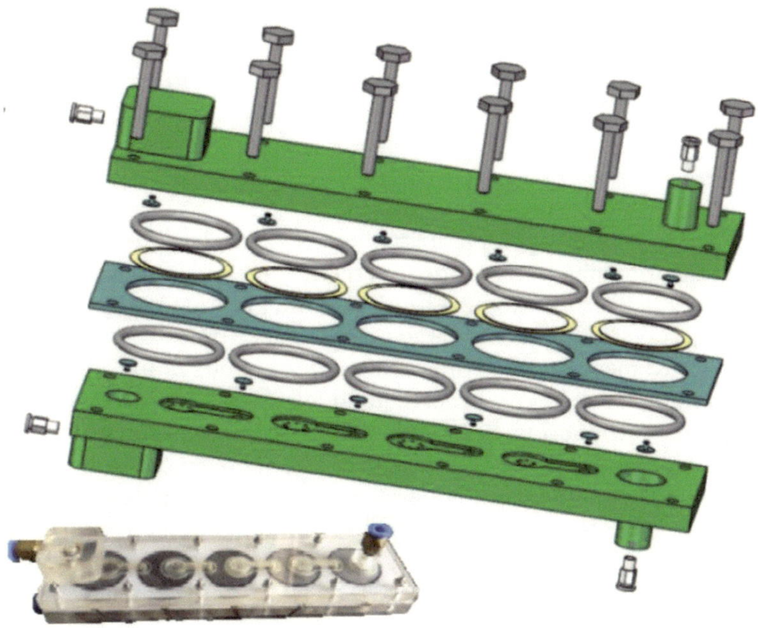

Figure 22.
The three-dimensional structure and prototype of PPFATC [17].

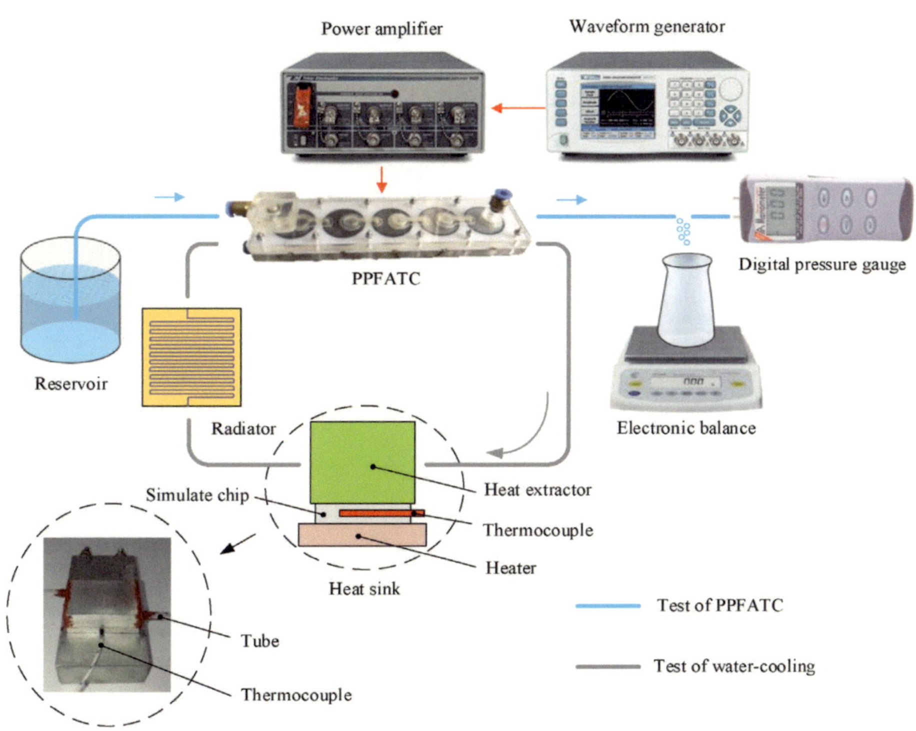

Figure 23.
The experimental platform of PPFATC and chip water-cooling system [17].

in **Figure 23**. The experiment shows that when the driving voltage is 60 V_{pp}, the maximum flow rate of the PPFATC in parallel is 251.1 ml/min and the maximum output pressure is 60.2 kPa. In series, the PPFATC can achieve 186.2 ml/min and

109.9 kPa, respectively. The water cooling system with PPFATC can obtain good cooling performance under low driving voltage. When the driving voltage is 60 V_{pp}, the PPFATC can reduce the chip temperature from 107.8–51°C.

4. Conclusion

Piezoelectric microfluidic technology is currently undergoing a period of prosperous development, partly motivated by the demands for pumping devices in the fields of drug delivery, biological fluid handling, micro total analysis systems, electrophoresis detection, liquid cooling of microelectronics and polymerase chain reaction (PCR). To meet various application requirements, quite a few novel principles and configurations have been presented over the last two decades, including liquid micropump, air micropump, single-chamber micropump, multi-chamber micropump, single-actuator micropump, multi-actuator micropump. This work mainly presented the structure and operating principle of single-chamber and multi-chamber piezoelectric micropumps and demonstrated an application in the field of chip water-cooling system. Some influencing factors on the performance of the piezoelectric micropumps were tested by the experimental frequency and voltage responses. Experimental results showed that there was an optimal driving frequency to maximize the flowrate of the piezoelectric micropump. Basically, the output flowrate was enhanced with the increasing driving voltage. It was helpful to increase the flowrate and backpressure of micropumps through combining the pumping chamber in parallel and serial, respectively. In this work, a maximum flowrate of about 280 ml/min could be achieved by combining five chamber in serial as well as the maximum output pressure approximately approached 11 kPa. Therefore, this work can be used as a reference and guideline for the design and application of piezoelectric microfluidic technology.

Author details

Song Chen*, Zhonghua Zhang, Junwu Kan, Jianping Li and Jianming Wen
Zhejiang Normal University, Jinhua, P.R. China

*Address all correspondence to: chensong@zjnu.cn

IntechOpen

© 2021 The Author(s). Licensee IntechOpen. This chapter is distributed under the terms of the Creative Commons Attribution License (http://creativecommons.org/licenses/by/3.0), which permits unrestricted use, distribution, and reproduction in any medium, provided the original work is properly cited.

References

[1] E. Kjeang, N. Djilali, D. Sinton, Microfluidic fuel cells: a review, J. Power Sources 186 (2009) 353-369.

[2] C.Y. Lee, C.L. Chang, Y.N. Wang, L.M. Fu, Microfluidic mixing: a review, Int. J. Mol. Sci. 12 (2011) 3263-3287.

[3] N.C. Tsai, C.Y. Sue, Review of MEMS-based drug delivery and dosing systems, Sens. Actuators A Phys. 134 (2007) 555-564.

[4] F. Amirouche, Y. Zhou, T. Johnson, Current micropump technologies and their biomedical applications, Microsys. Technol. 15 (2009) 647-666.

[5] D.J. Laser, J.G. Santiago, A review of micropumps, J. Micromech. Microeng. 14 (2004) R35-R64.

[6] B.D. Iverson, S.V. Garimella, Recent advances in microscale pumping technologies: a review and evaluation, Microfluid. Nanofluid. 5 (2008) 145-174.

[7] K.S. Lee, B. Kim, M.A. Shannon, An electrostatically driven valve-less peristaltic micropump with a stepwise chamber, Sens. Actuators A: Phys. 187 (2012) 183-189.

[8] Z.R. Xu, C.G. Yang, C.H. Liu, Z. Zhou, J. Fang, J.H. Wang, An osmotic micro-pump integrated on a microfluidic chip for perfusion cell culture, Talanta 80 (2010) 1088-1093.

[9] J. Huang, J.H. Zhang, S.Y. Wang, Theory and experimental verification on valveless piezoelectric pump with multistage Y-shape tubes, Opt. Precis. Eng. 21 (2013) 423-430.

[10] D.N.C. Nam, K.K. Ahn, Design of an IPMC diaphragm for micropump application, Sens. Actuators A: Phys. 187 (2012) 174-182.

[11] H.H. Kim, J.H. Oh, J.N. Lim, K.J. Lim, D.H. Park, Design of valveless type piezoelectric pump for micro-fluid devices, Proc. Chem. 1 (2009) 353-356.

[12] Zhang Z, Kan J, Cheng G, et al. A piezoelectric micropump with an integrated sensor based on space-division multiplexing [J]. Sensors and Actuators A: Physical, 2013, 203(29-36).

[13] Zhonghua Zhang, Jun wu Kan, Shuyun Wang, Hongyun Wang, Jijie Ma and Yonghua Jiang. Development of a self-sensing piezoelectric pump with a bimorph transducer. Journal of Intelligent Material Systems and Structures, 2016, Vol. 27(5) 581-591.

[14] Zhonghua Zhang, Junwu Kann , Shuyun Wang, Hongyun Wang, Jianming Wen, Zehui Ma. Flow rate self-sensing of a pump with double piezoelectric actuators. Mechanical Systems and Signal Processing 41 (2013) 639-648.

[15] Chen S, Yu M, Kan J, et al. A Dual-Chamber Serial-Parallel Piezoelectric Pump with an Integrated Sensor for Flow Rate Measurement [J]. Sensors (Basel), 2019, 19(6).

[16] Zhang Z, Chen S, Wang S, Kan J, Wen J, Yang C. Performance evaluation and comparison of a serial–parallel hybrid multichamber piezoelectric pump. Journal of Intelligent Material Systems and Structures. 2018; 29(9):1995-2007.

[17] Lu S, Yu M, Chen S, et al., A Quintuple-Bimorph Tenfold-Chamber Piezoelectric Pump Used in Water-Cooling System of Electronic Chip. IEEE Access, 2020(8): 186691-186698.

Chapter 8

The Roles of Piezoelectric Ultrasonic Motors in Industry 4.0 Era: Opportunities & Challenges

Sahil P. Wankhede and Tian-Bing Xu

Abstract

Piezoelectric Ultrasonic motors (USM) are based on the principle of converse piezoelectric effect i.e., vibrations occur when an electrical field is applied to piezoelectric materials. USMs have been studied several decades for their advantages over traditional electromagnetic motors. Despite having many advantages, they have several challenges too. Recently many researchers have started focusing on Industry 4.0 or Fourth Industrial revolution phase of the industry which mostly emphasis on digitization & interconnection of the entities throughout the life cycle of the product in an industrial network to get the best possible output. Industry 4.0 utilizes various advanced tools for carrying out the nexus between the entities & bringing up them on digital platform. The studies of the role of USMs in Industry 4.0 scenario has never been done till now & this article fills that gap by analyzing the piezoelectric ultrasonic motors in depth & breadth in the background of Industry 4.0. This article delivers the novel working principle, illustrates examples for effective utilization of USMs, so that it can buttress the growth of Industry 4.0 Era & on the other hand it also analyses the key Industry 4.0 enabling technologies to improve the performance of the USMs.

Keywords: piezoelectric ultrasonic motors, industry 4.0

1. Introduction

Increasing functionalities and weights/sizes reductions are critical issues for future aircrafts, space exploration vehicles, space instrumentations and industrial application etc. One challenge is the miniaturization of motors, wherein the efficiency of commonly used electromagnetic coil-based motors is dramatically reduced when their size less than centimeters scales. On the other hands, the rapid developments on piezoelectric ultrasonic motors (USMs) may fill the technical gap. Piezoelectric ultrasonic motors have been used in various technological fields from past decade in the gadgets, which we are using in our daily life i.e., a mobile phone to most advanced applications in aerospace. Recent advanced sciences & technologies developments on complex & tech-savvy products like satellites, mobile phones, camera lens, spaceships, automotive, robotics, biomedical instruments, manufacturing, etc., makes our life more convenient. These new products have raised new demands on modern motors like micro scale, light weight, high torque, no electromagnetic interference, low noise etc., which cannot be met by traditional motors. To make bridge for this gap, many scientists developed specialized motors, such as electrostatic motors, USMs, bionic motors, photo-thermal motors, shape memory alloy motors, microwave motors, etc. Among them, USMs have more advantages [1].

Although the first concept of USMs was invented in 1948, when just after the World War II, USMs have been used for practical applications in 1980s. It works on the principle of converse piezo electric effect *i.e.*, vibrations occur when an electrical field is applied to some piezoelectric structures. Similar to traditional electromagnetic motors this kind of motors comprises of stator & rotor, however, the difference is that a USM consists of piezoelectric structure, which is bonded to stator instead of coil and magnet pairs to make simpler and compact in size. In USMs, the piezoelectric structure is first vibrated in ultrasonic frequency band (>20 kHz), which in turn vibrates the stator when a driving voltage is applied in matched frequency. Thus, the frictional contact force between the stator & rotor or slider leads to the mechanical movement & torque. USMs can obtain high torque/weight ratio (torque density) in comparison with traditional electromagnetic motors because they are compact in structure & flexible in design. They are capable to drive the payloads directly without connecting to gear or gear train mechanism for some special applications. Most important that USMs quickly response commends in less than a few microseconds due to the advantages of piezoelectric materials and small inertia of rotors. Furthermore, they have capability of self-locking, high holding torque, precision motion control which can be utilized for application areas requiring high degree of precise motion for, e.g., medical operations & manufacturing or inspection of intricate products. In addition, USMs have zero electromagnetic interference which is one of the prominent applications in magnetic resonance imaging (MRI). USMs can also be operated in extreme temperature conditions which makes them first choice to use in aerospace application for e.g., space mission. Apart from above, they are silent during their operating cycle which makes them suitable for low noise applications [1].

Although USMs has many advantages, there are still several challenges remaining to achieve them in key areas like new design, motion control, piezoelectric material, friction & wear, thermal performance, modeling & optimization and advanced manufacturing technologies [2]. These challenges needed to be addressed in order to make USMs to be broadly utilized with its full potential in modern day industrial settings & diverse field of applications.

Fourth Industrial revolution or Industry 4.0 is the current phase of the industry wherein its emphasis on the digitization & interconnection of the products & services throughout the product life cycle i.e., from the birth of the product to the end of the life of product/services. Industry 4.0 relies mainly on various technological tools, which include but not limited to Big Data & AI analytics, Augmented reality (AR) Additive Manufacturing, Cyber Security, Industrial Internet of Things (IIoT), Autonomous Robots, Digital Twins, Horizontal & Vertical integrations & Cloud computing for carrying out the process of digitization & interconnection. Therefore, in this article, we not only take advantage of Industry 4.0 tools to improve the performance of the USMs but also promote the applications of USMs that can make them best fit into Industry 4.0 settings.

This chapter is divided into five sections. In the first section, we will introduce the history of development of USMs, types of USMs, structure & operating mechanism of USMs, and piezoelectric materials used. In the second section, the reviews of various articles, especially the publications in the last 5 years on ultrasonic motors from eclectic sources i.e., conference proceedings, journals, US patents & doctoral thesis. It analyses them & gives a brief bibliographic summary comprising of publication year, journal of publication, country of origin of research. The classification of USM technologies in different categories, which include mainly new design, motion control, piezoelectric material, friction & wear, thermal performance, modeling & optimization & provide comprehensive summary of the articles describing achievements, challenges & opportunities, will be presented in the third section. The fourth Section of this chapter will briefly introduce industry 4.0 & key enabling technologies, i.e., Big Data & Artificial Intelligence analytics, Augmented Reality, Additive manufacturing, Industrial internet of things (IIoT), Digital Twins & simulation. In the Fifth section, it will be addressed that the approaches to improve the overall performance of piezoelectric motors by effective utilizing key enabling technologies offered in Industry 4.0 settings. It further elaborates on the various types & applications of piezoelectric motors that can be utilized effectively to foster Industry 4.0 expectations.

2. Piezoelectric ultrasonic motors

Development of Piezoelectric USMs has been in progress since 1980. Piezoelectric motors worked on the principle of reverse piezoelectric effect i.e., electrical energy applied to piezoelectric substrate is converted into mechanical actuation or motion in this case it refers to vibration. These motors are known as "Ultrasonic Motors" since the frequency of vibration of the piezoelectric element inside the motor is in the range of ultrasonic frequency band i.e., greater than 20 kHz. The chronology of events in the development of ultrasonic motors is summarized in **Table 1** [1].

2.1 Classification

Ultrasonic Motors do not have a uniform methodology for classification because of the design flexibility & structural diversity [1]. For application point of view, USMs can be classified as rotary and linear type motors; from vibration shape, USMs can be classified as rod shape, π-shape, ring shape, and cylinder shape, from vibration characteristics, USMs can be classified as standing wave and propagating wave types; etc. More details of classification are done by viewing angle as illustrated in **Table 2** [1]. Classification is also done based upon the vibration type i.e., Longitudinal Vibration, Longitudinal Bending Vibration, Longitudinal-torsional vibration, Bending Vibrations & In-Plane vibrations [1].

2.2 Operating mechanism of USM

The most common USMs is the Traveling Wave Ultrasonic Motors (TRUMs) because of their simple constructions and broad applications. One typical TRUM is mainly composed with a stator, rotor, shell, bearing, spring, friction linear, PZT, base, etc. Piezoelectric ceramic is affixed to stator while rotor is affixed by friction liner [1]. Friction liner is bonded to a rotor, which contacts with stator through axial pressure. The traveling wave is formed by superposition of two mode responses with equal amplitude and phase difference $\pi/2$ both in time and space. If pre-pressure is applied to the rotor, then the vibration with micro amplitude of points on stator surface will be transformed to rotary motion of the rotor through frictional force. A structural diagram of TRUM is shown in **Figure 1** and the working principle of traveling wave, which is formed inside of the USM leading to the motion of the rotor, is presented in **Figure 2**. For successful development of the traveling wave, it is necessary that the two resonant modes with an identical frequency and mode shapes (standing waves) in an elastic body have $\pi/2$ phase difference both in

Year	Name of Scientist/ organization	Chronology of Events in the development of USM
1948	Williams and Brown	First patent of "piezo motor" was applied.
1965	Lavrinenko	Patent invention was granted for using piezoelectric plate to drive rotor.
1973	Barth from IBM	Two piezoelectric actuators to produce longitudinal vibration of horns was used. The rotor is driven by the contact friction between the rotor surface and end of the horns.
1975	Vishnevsky	The edge of a rectangular piezoelectric composite stator was pressed by the spring, which excited a longitudinal vibration mode to drive the rotor
1981	Lithuanian Vasiliev	An ultrasonic motor with the ability of driving larger loads of gramophone wheel. It became the first practical application of piezoelectric actuator.
1982	Sashida	A standing wave USM was designed. Piezoelectric ultrasonic motor met the performance requirements for actual applications for the first time.
1983	Sashida	Traveling wave USM was designed.
1985	Kumada	A longitudinal-torsional hybrid ultrasonic motor driven by single phase signal.
1987	Ishc from Panasonic, Inc	A ring type traveling wave ultrasonic motor based on Sashida's traveling wave motor.
1987	Canon Co.Ltd	The ring-type ultrasonic motor in the zoom lens of EOS camera was used first time in Engineering application
1995	Zhao, Chunsheng	A disk-type traveling wave ultrasonic motor

Table 1.
Summary of the development of ultrasonic motors [1].

Viewing Angle	Type
Wave propagation method	Traveling Wave, Standing Wave
Movement Output way	Rotational, Linear
Contact State between the stator & rotor	Contact, Non-contact
Excitation conditions of stator by piezoelectric components	Resonant, Non-Resonant
Number of degrees of freedom of the rotor	Single degree of freedom, multi-degree of freedom
Displacement of operating mode in direction	Out plane, In-plane
Geometric shape of stators	Disks, Ring, Bar & Shell
Rotary directions	Unidirectional, Bidirectional

Table 2.
The classification of UTM methodologies [1].

space and time [1] & quarter wavelength difference i.e., $\lambda/4$ [2]. Consider a point P on the stator the traveling process of point P is illustrated in **Figure 2a–d**. At time t = 0 point P is at initial position as shown in **Figure 2a**, at t = T/4, **Figure 2b** the wave propagates right the wave peak reaches point P further at t = T/2, **Figure 2c** wave & Point P moves $\lambda/4$ forward then at t = 3 T/4 **Figure 2d** wave valley reaches point P & at t = T point P reaches initial position [1].

Figure 1.
Structural diagram of TRUM [1].

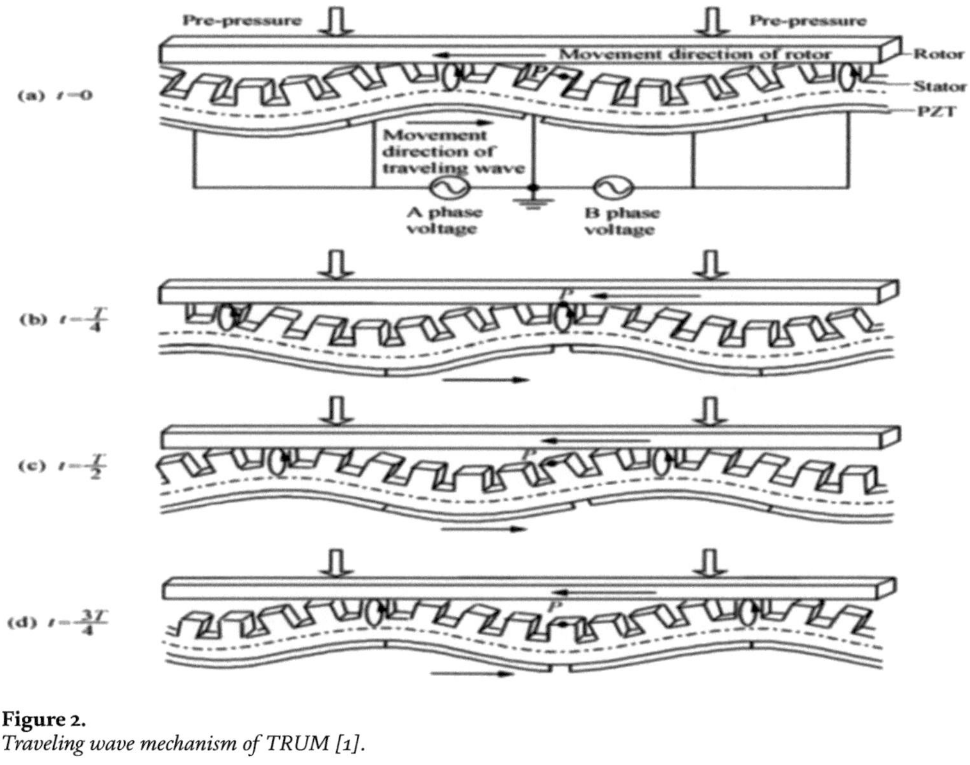

Figure 2.
Traveling wave mechanism of TRUM [1].

In order to achieve traveling wave, selection of right piezoelectric materials is required. Most commonly used piezoelectric materials are Barium Titanate BaTiO3, Lead Zirconate-lead titanate (PZT), relaxor ferroelectrics of $Pb(Mg_xNb_{1-x}O_3)$-$PbTiO_3$(PMN-PT) and $Pb(Zn_xNb_{1-x})O_3$-$PbTiO_3$(PZN-PT) single crystals [1].

3. Review on piezoelectric ultrasonic motors

This section provides a brief summary of the articles published since last 5 years since 2015 [2–298]. The articles are taken from eclectic resources like conference proceedings, high impact journals, patents & masters, Ph.D. thesis from reputed universities. After reviewing the articles, we found that most of the articles on ultrasonic motors found were stressing upon research areas like new design, modeling & optimization, motion control, friction & wear, piezoelectric materials used in USM, thermal performance, applications of USM & review papers. [2] We carried out statistical analysis for the research articles published on USM during this period. We plotted graph for number of publications with research areas (**Figure 3**), with year of publication (**Figure 4**), by country of research (**Figure 5**) & journal publication (**Figure 6**). From **Figure 3** one can see that majority of the articles published on USM stressed upon new design & modeling & optimization. **Figure 4** shows that gradual increase in the trend of number of publications. **Figure 5** indicates that countries like China & Japan are the leaders in doing research on USM. **Figure 6** shows journal "Ultrasonics" & "Sensors & Actuators A: Physical are the favorite among researchers & publishers for publishing articles on USM. In the following section, we are going

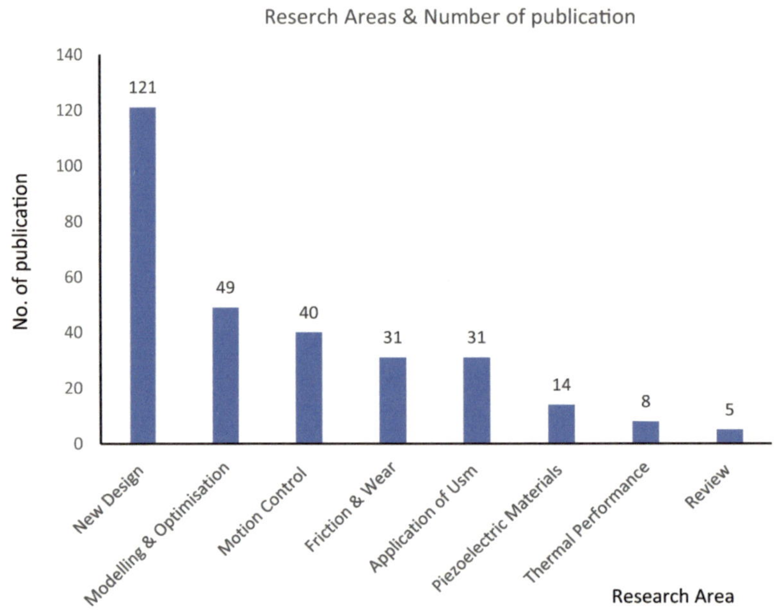

Figure 3.
Summary of number of publications in each research areas of USM in the years of 2015–2020.

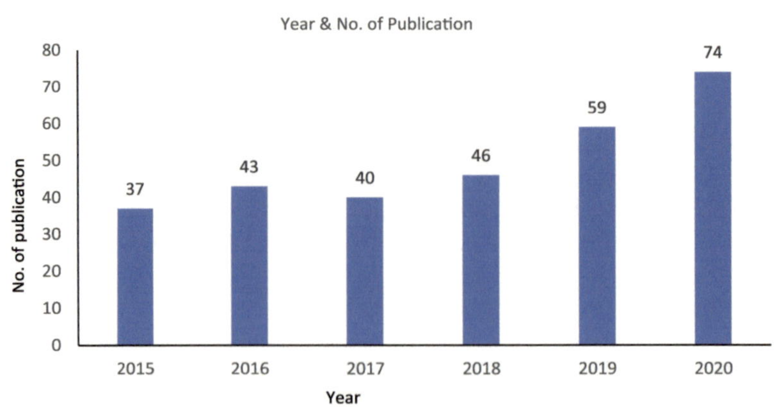

Figure 4.
Summary of number publications per year on USM from 2015 to 2020.

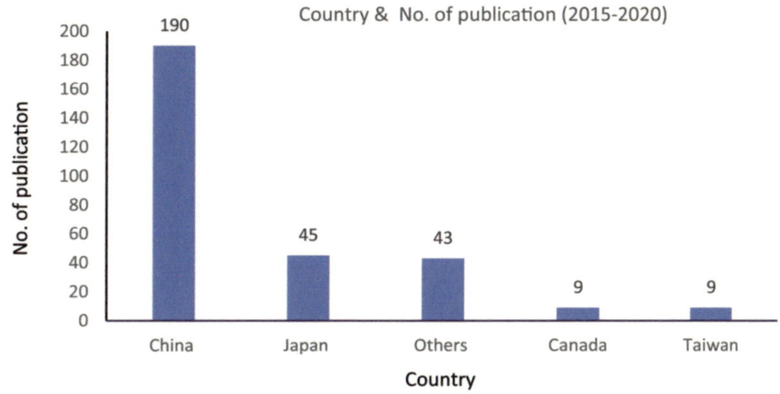

Figure 5.
Summary of USM publication numbers of author's country and region distribution.

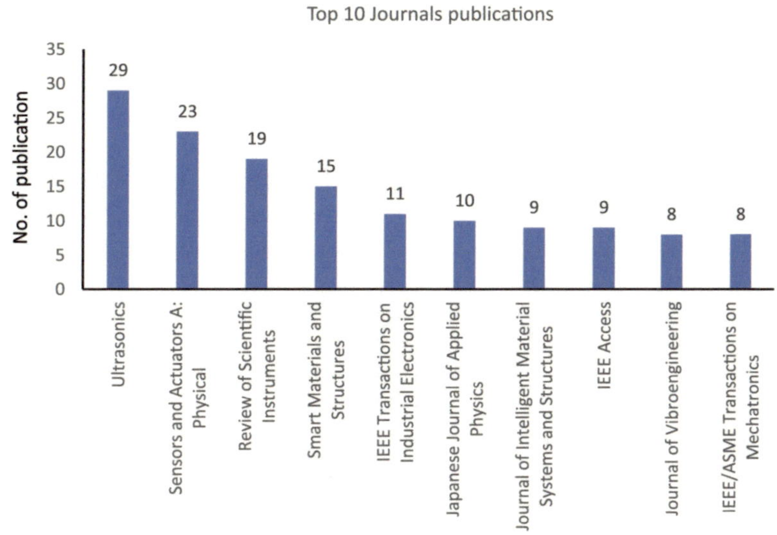

Figure 6.
Summary of USM publication numbers in the top ten journals.

to briefly discuss the achievements, challenges, & opportunities about those publications made in the areas identified above.

3.1 New design

In this section, we summarize the novel ideas proposed by the researchers for designing & developing ultrasonic motors as well as optimizing their designs in order to improve their efficiency & performance. The major objectives are to make UTMs with small size, high torque, and high-power density. Among them, miniaturization of the motor is the key challenge to achieve.

3.1.1 New developments for miniaturization

Since 2015 various research articles emphasized upon miniaturization, micro USM, scaling to sub millimeter range. Tomoaki Mashimo et al., a research group at Japan, develop serial of micro ultrasonic motor (μ-UTMs) with volume scale of a few cubic millimeters to submillimeter [3–10]. Those μ-UTMs include rotary and linear types for different applications. Don L. DeVoe et al. developed one of the smallest bulk PZT TRUMs capable of bidirectional motion with PZT stator of diameter 4.12 mmand 323 mN preload force. The motor stator was fabricated using micro powder blasting of homogeneous PZT sheet. It achieved a maximum speed of 30 rpm & stall torque of 501 mN-mm [11]. Yingxiang Liu et al. carried out an overall weight of 8.5 g longitudinal–bending hybrid linear USM, whichis able to achieve 487 mm/s no-load speed, the maximum output force of 2.3 N, and & weight of the prototype obtained, and respectively [12]. Qiquan Quan et al. developed U shaped piezoelectric ultrasonic motor that mainly focused on miniaturization and high-power density [13]. Fulin Wang et al. developed a miniature spherical ultrasonic motor using wire stators for directional adjustment of a vascular endoscopic camera [14]. Ho et al., proposed a miniaturized simple shear vibration piezoelectric screw-driven structure USM to drive the high precision linear motor [15]. Zhou et al., developed a novel a radius of 2 mm three-dimensional contact model of piezoelectric TRUM utilizing MEMS fabrication technology [16].

3.1.2 New developments for high power densities

In addition to miniaturization some researchers focused on high torque and power density aspects of piezo motors, Mizuno et al., developed a hybrid torsional/bending (T/B) modes USM to provide high driving force, large driving distance, and low weight, resulting into high torque density and high-power density [17]. It is constructured with rod-shaped transducer operating in torsional/bending (T/B) modes and excited two elliptical motions on its bilateral ends to drive the rotor orthogonally pressed onto the transducer. Chang et al., developed a ring-shaped traveling wave ultrasonic motor with a suspension stator for improving output power density [17]. The maximum stator vibration amplitude of 4.25 μm, which is nearly 4.7 times of that without suspension, with speed of 62 rpm and a stall torque of 49.5 mNm was observed, under a driving signal of 30 V_{pp} when the mass block was 0.30 g. Fan et al. developed a miniaturized ultrasonic motor with a high thrust–weight ratio by using the first order bending vibration mode (B1 mode) and second order bending vibration mode (B2 mode) to realize bidirectional movement through a single-phase driving signal [19]. Li et al., constructed screw-type USM with a three-wavelength exciting mode to achieve a high-output thrust [20].

In order to meet the requirement of large thrust & maximum output few research articles on the ultrasonic motors were based upon typical shapes, such as U-shape, V-shape, L-shaped, and Π-type. For instance, the structure of the linear ultrasonic motor with a laminated stator, which was made of two identically single U-shaped stators, was proposed by Sun et al., [36]. The testing results showed that the maximum output force of the laminated motors increases by 40% than that of single layer U-shaped motor, while the maximum velocity increased by 38%. Yao et al. proposed a novel large thrust-weight ratio V-shaped linear USM with a flexible joint operated in the coupled longitudinal-bending mode. The motor had a compact size and a simple structure with a large thrust-weight ratio (0.75 N/g) [37]. Furthermore, they also proposed a novel large thrust L-shaped linear USM utilizing the antisymmetric and symmetric modes of the L-shaped stator operating in a single resonance mode to realize the bidirectional motion of the slider [38]. In order to meet the demand of a linear ultrasonic motor with large thrust in narrow space, a novel Π-type linear ultrasonic motor with double driving feet was constructed [39]. The motor had structural stability and high dynamic performance, such as a no-load speed of 273 mm/s and 238 mm/s in two directions, corresponding to a maximal thrust of 80 N and 110 N. Wang et al., constructed a V-type motor having two driving feet and a simple structure, which torque applied to the motor was converted into a normal preload between the driving feet and the mover to avoid the use of a large preloading mechanism [40]. The maximum no-load velocities of the motor moving to the right and left are 85.2 mm/s and 76 mm/s, respectively, and the maximum output force is 1.96 N.

Light weight, high torque & desired output performance are some of the important features required in USMs. Niu et al., developed a light arch shaped, four legged linear & hollow USMs, which a light arc-shaped USM with the first-order longitudinal vibration mode and the second order bending vibration mode were superimposed in the stator plane, in order to meet the requirements [45]. The output torque of the USM under the single-stator configuration reached up to 2.6×10^{-2} N·m, and the double-stator which was 1.5 times greater than that under of single stator configuration.

To realize applications involving low speed and high torque in the high-performance actuator industry, especially in the aerospace field, a novel 70H (Hollow) TRUM with an outer diameter of 70 mm and an aperture ratio of 53% (the ratio between the aperture and outer diameter) with a mass of 210 g was developed [46]. The TRUM, The torque density of 11.43 N·m/kg, maximum no-load speed of 50 rpm, and the maximum stall torque 2.4 N·m were achieved.

The influence of the vibration mode of the stator and the structural dimensions of the metal elastomer and piezoelectric ceramic ring on the effective electromechanical coupling coefficient (EMCC) was analyzed by Niu et al. [47]. The efficiency of a hollow USM was improved by optimizing the stator's effective electromechanical coupling coefficient. In addition, a four-legged linear ultrasonic motor with a new structure which is the in-plane first-order longitudinal vibration mode and the out-of-plane anti-symmetric vibration mode superimposed to produce linear motion [48]. The USM consists of a stator and four groups of eight piezoelectric ceramic sheets. The experimental results for a prototype 600 × 160 mm showed the maximum translational speed could reach 135 mm/s and the maximum thrust of 3.6 N with a 200 V driving voltage. The USM had the advantages of simple structure and high output efficiency, which made it suitable for precision systems and industrial applications.

Izuhara et al. proposed a linear piezoelectric motor using a hollow rectangular stator that can translate a load placed inside it by a direct drive [49]. This stator structure enabled a quick response and high resolution by few components for controlling autofocus and zoom mechanisms in imaging devices.

3.1.3 Multiple degrees of freedom piezoelectric ultrasonic motor (multi-DOF-USM)

Shi et al., constructed a new type of multiple-degree-of-freedom (Multi-DOF) compact structure USM to achieve high output torque [33]. It consisted of a ring type composite stator with four driving feet uniformly arranged in the inner circumference of the ring stator. The stator employs two orthogonal axial bending modes and a radial bending mode, by exciting two of them simultaneously, to generate elliptic trajectories on driving feet tips and to push sphere rotor around x, y and z axis respectively. Su et al. improved the performance of a non-resonant piezoelectric motor, which is a symmetric piezoelectric linear motor driven by three-phase square-triangular waves signal and four-phase sine waves signal of peak to peak value 100 V at 100 Hz with 50 V offset [34]. The speeds of prototype reached 733 μm/s and 667 μm/s and the maximum thrust is 8.34 N and 6.31 N respectively. Similarly, a non-resonant linear ultrasonic motor utilizing longitudinal traveling waves was proposed by Liang Wang et al. [35]. The stator system was modeled by utilizing the transfer matrix method (TMM). The motor prototype achieved a maximum mean velocity of 115 mm/s and a maximum load of 0.25 N.

Li et al. proposed electromagnetic-piezoelectric hybrid driven three-degree-of-Freedom USM which is hybrid driven electromagnetic filed and electrical field [41]. In one of their design, a novel ball-type spherical multi-DOF USM, composed of three built-in stators and a hollow spherical rotor was developed and tested for the design of a compact multi-degree-of-freedom (multi-DOF) piezoelectric driven actuator [42]. The rotational speeds of X-axis, Y-axis and Z-axis can reach 29 r/min, 17 r/min and 16 r/min, respectively, when the frequency matches, which verifies the feasibility and rationality of the multi-DOF movement of the motor. They also proposed a multi-DOF spherical USM with built-in traveling wave stators, in which each traveling wave stator could be controlled independently and the spatial arrangement of the support structures [42]. The maximum speed achieved 45.6 rad/min with output torque of 1.265 Nm when an excitation voltage of 400 V with the preload of 100 N. The motor had the advantages of large output force and adjustable preload.

Kazokaitis et al., developed a novel design of a multi-DOF USM, which is combined the magnetic sphere type rotor and two oppositely placed ring-shaped piezoelectric actuators into one mechanism [44]. Such a structure increases impact force and allows rotation of the sphere with higher torque useful for attitude control systems used in small satellites.

3.1.4 Preload effect study

Contact mechanism between the stator & rotor is one of the important factors responsible for the efficient performance of the USMs. The studies of contact surface, contact mechanism, preloading method of USMs has been one of the prominent topics in the USM research field. Zhang et al., proposed a solution to reduce the radial sliding by optimizing the stator comb-teeth of a TRUM [50]. They further developed a 3D finite element model for longitudinal torsional USM by ADINA in order to study the mechanical simulation and contact analyses [51]. A novel hollow type USM, which the preload was applied from the bottom of the stator through a wave spring, was proposed, [52] It could not only enhance the anti-overload ability but also extended the working life of the motor.

Wang et al., analyzed the characteristics of a TRUM with considering the structural stiffness of the preload structure [53]. It demonstrated that the pre-pressure on the rotor was not a constant value because of the structural stiffness of the preload structure. In addition, it explained the driving mechanism of the TRUM under unsteady pre-pressure and deduced a dynamic model considering the stiffness of the preload structure.

In addition, contact force analysis by Hertz contact theory popped up in few research articles. Dong et al., carried out design and performance analysis of a TRUM with double vibrators [54]. The analytical model of double-vibrator motor was established based on elliptical distribution rule of surface point velocity, linear superposition of motions and contact force analysis under Hertz contact theory. Pan et al., focused on the coupling relationship between the flywheel vibration and the gimbal rotation through the variable stiffness of the bearing [54].

3.1.5 Multivibrators

Some research articles emphasized on novel idea of constructing USMs by using multivibrators. Yang et al. illustrated that TRUM with double vibrators can improve the output performance effectively [56]. Inheriting the concept of two traveling waves propagating in the stator and rotor, a dual traveling wave rotary ultrasonic motor (DTRUM) energized only in the stator was proposed. The experimental results showed that the performance of dual traveling wave TRUM was superior to the TRUM with single traveling wave. The no load speed was 60 rpm and the stalling torque was 0.85 Nm. They further presented, an optimal design of a double-vibrator USM using combination methods of finite element method, sensitivity analysis and adaptive genetic algorithm [57]. The measured results showed that this method was effective for the optimal design of ultrasonic motors. Lu et al. proposed a new idea for constructing the motor with the stator containing several vibrators fabricated by bonding piezoelectric ceramics (PZTs) to a metal base [58]. The longitudinal and bending modes were excited in the vibrators by two alternating current (AC) voltages with a 90° phase difference were applied. The bending vibrations of the vibrators were stacked to form the torsional vibration of the stator, ultimately generating longitudinal-torsional composite vibration. Mohammed & Zakariyya proposed an idea on the development of a new type of a linear USM with double cantilever vibrators [59]. The resonance frequencies of the vibrators were 21.33 kHz, and this was also the frequency in which the two vibrators were driven to determine the output parameter such as driving force and velocity.

Multi-vibration mode USMs and sandwich type USMs were designed & analyzed in some of the research articles. Zhou et al., developed a novel multi-mode differential USM with two sandwich-type transducers, which utilized the diverse combination of four vibration modes: symmetrical and anti-symmetrical first longitudinal modes, symmetrical and antisymmetric second bending modes, which it could realize three-step speed regulation with different speed-thrust force characteristics by switching the operation mode [65]. They also presented a novel 2-DOF planar linear USM which the stator of the motor was divided as two transducers and two isosceles triangular beams [66]. The operating principle of the USM and the formation of the elliptical trajectory of the driving foot were analyzed, and the variable mode excitation method was illustrated. This motor can gain a maximum speed of 211.3 mm/s with thrust force of 3.15 N under an exciting voltage of 400 V_{P-P}. A new sandwich type ultrasonic motor using combination of the first symmetrical and anti-symmetrical longitudinal modes was presented by them [67]. The working principle of the motor and the elliptical trajectory formation of the driving foot were analyzed. A new linear USM using hybrid mode of the first symmetric and anti-symmetric longitudinal modes was described [68]. The stator was constructed by two Langevin transducers in combination with two isosceles triangular beams. Zhou et al., constructed a rotary USM with rotationally symmetrical structure, which the stator consists of four connected sandwich-type transducers and eight driving feet [69]. With the driving frequency of 50.93 kHz and voltage 300 V_{P-P}, the motor gave a maximal no-load speed of 157.9 r/min and a maximal output torque of 11.76 mNm.

Lu et al., proposed a single-modal linear motor based on multi vibration modes which contained two kinds of PZT ceramics [70]. The linear motor works by exciting the transverse vibration mode of the PZT ceramic on the upper surface of stator elastomer and the shear vibration mode of PZT ceramics at two ends simultaneously. The no-load velocity and the maximum output force reach 169.4 mm/s and 1.1 N, respectively. Mizuno et al., developed a high-torque sandwich-type MDOF-Spherical USM using a new annular vibrating stator with a strong excitation structure [70]. The maximum torques of rotation around the X(Y)-axis and Z-axis were measured as 1.48 N·m and 2.05 N·m respectively. Moreover, the values for torque per unit weight of the stator were obtained as 0.87 N·m/kg for the X(Y)-axis and 1.20 N·m/kg for the Z-axis, separately. Ma et al., developed a compact motor in which the stator composed of two piezoelectric plates attached to a T-shaped steel body [72]. Two orthogonal bending modes were excited by driving one piezoelectric plate and the reversed motion of the rotor could be obtained by driving the piezoelectric plate on the opposite side. Maximum power of 2.3 mW and efficiency of 9% with a load of 0.8 mN m at a rotation speed of 27 rpm were obtained for a prototype stator with a size of 15 mm × 2.44 mm × 2 mm, operated at 44.8 kHz.

Ceponis et al., presented a numerical and experimental investigations of a multimodal TRUM, which is driven by four electric signals with phase difference of $\pi/2$., being able to generate up to 115 RPM rotation speed at constant preload force [81]. They further proposed a new flat cross-shaped USM, which operation principle based on the first in-plane bending mode of the cross-shaped stators driven by four harmonic signals with phase difference of $\pi/2$ [82]. The advantages of the motor were high rotation speed, simple and scalable design, and the small space required for motor mounting wherein it can be directly mounted on the printed circuit board. Prototype achieved a maximum rotation speed of 972.62 RPM at 200 Vp-p when the preload force of 22.65 mN was applied.

Tanoue et al., designed a novel ultrasonic linear motor equipped with a quadruped stator that used the first longitudinal mode and the first and second bending modes [85]. A maximum driving speed of 148 mm/s and a maximum thrust of 294 mN were achieved for a device with a total length of 20 mm and a weight of 5 g. One more linear USM that drives a slider rod inside the quadruped stator to realize a compact linear motion system was proposed by them [86]. Maximum no-load speed of 258 mm s-1 and maximum thrust of 490 mN were obtained with total length of the stator transducer of 20 mm and its weight of 4.9 g. Cheon et al., proposed a new type of ultrasonic rotary motor that could replace existing ultrasonic motors for driving camera zoom lenses and investigated experimentally [87]. Peng et al., presented a new kind of the rotary USM with a longitudinal vibration model of the Langevin transducer acting as the stator, while the rotor consisted of a shaft and spiral fins, the spiral fins working as an elastic coupling component by which it cannot change its direction because the spiral fins' incline direction was fixed [88]. This motor can be used when one directional motion was required. Romlay et al., proposed an improved stator design of TWUSM using the comb-teeth structure which was expected to increase the overall efficiency [89]. Le et al., proposed a novel design methodology to optimize actuator configuration for linear USMs by considering the dynamic behavior of the stator in its operating environment, where it interacts mechanically with the moving stage and other peripheral components [90]. This helped to evaluate the actuator output performance parameters for design optimization. Pan et al., developed a novel low-friction type piezoelectric rotary motor based on centrifugal force with high speed, high power, and high efficiency output, novel low-friction type piezoelectric rotary motor [91]. Yang et al., proposed a dual-rotor hybrid USM with four side panels without using the torsional piezoelectric ceramics, which was indirectly excited by four uniformly distributed side panels along the circumference of stator cylinder [92]. The stalling torque of the prototype is 8 mNm and the no-load speed is 140 r/min was obtained at 44.7 kHz for a prototype with the size 27.2 mm x 27.2 mm x 70 mm, while the outer diameter of the stator cylinder was 20 mm. The experimental results indicate that the motor could operate in the first longitudinal and the second torsional coupled vibration modes transformed from the first longitudinal and the first bending vibration modes of four side panels.

Jiu et al., proposed a modal independent USM with dual stator based on optimizing the location of a rotor and two stators which excited at the same mode [93]. Modal test showed the disparity between the modal frequencies of the stators was 0.78%. The rotary speed of the USM is 75 revolutions per minute (clockwise) and 65.8 revolutions per minute (anti-clockwise) with the maximum torque of 8.4 N. mm at the voltage of 400 V_{p-p}. Li et al., proposed a traveling wave ultrasonic motor with a metal/polymer-matrix material compound stator which the stator was composed of a metal ring and polymer-matrix teeth [94]. The main merits of the proposed ultrasonic motor were low cost, light weight, high processing efficiency and long life. Sanikhani proposed a new linear ultrasonic motor based on the orthogonal vibration modes of an elliptical shaped [94]. Based on the experimental results, the prototype has a no-load speed of 40 mm/s and maximum thrust force of 1.55 N under excitation voltage of 70 Vp and preload of 12 N. Sun et al., developed a novel cylindrical ultrasonic motor easy to be fixed [96]. Two orthogonal B_{03} bending vibration modes of the stator were generated with temporal shift of 90 to produce elliptical movement on the driving surface. The weight of the proposed stator and motor was only 2.56 and 4.1 g, respectively & it achieved a maximum speed of 170 r/min under working frequency of 31.6 kHz [96].

In order to reduce the driving voltage and gain better output characteristics of piezoelectric actuators, an eight-zonal piezoelectric tube-type threaded ultrasonic motor based on two second order bending modes was analyzed by Chu et al. [97]. The USM could output a stall force of about 5.0 N and a linear velocity of 4.9 mm/s with no load at the driving voltage of 40 V_{pp}. This USM with a compact structure and screw drive mechanism showed fine velocity controllability and had great application in micro-positioning systems. Borodinas et al., described a USM that used the radial mode of excitation of the double ring's stator [98]. The main goal of the proposed design was to increase motor performance using d_{33} ceramic polarization working in the radial mode. The motor could be driven by a simple harmonic signal and used for standard piezoceramic rings [98]. An ultrasonic linear motor with dual piezoelectric (PZT) actuators which a traveling wave motion was generated on the stator by a double-sided excitation of the stator of the USM, was developed by Yang et al. [99]. The simulation results showed the differences to the characteristics that are achieved by adjusting the critical parameters, such as the PZT boned positions, the excitation frequency and the preload, in order to derive the best design [99]. Aoyagi et al., analyzed an application of noncontact transportation utilizing the near-field acoustic levitation phenomenon, which is a rotary-type noncontact-synchronous ultrasonic motor using acoustic viscous force [100]. Xu et al., proposed a novel rotary ultrasonic motor with two longitudinal transducers [101]. Only first order longitudinal vibration mode was used in the ultrasonic motor, which avoided the frequency degeneration of modal coupling ultrasonic motors. Mechanical performances showed that the motor can obtain rotary speed of 350 r/min and the maximum torque is 186 N·mm under the voltage of 300 Vp − p. Wu et al., fabricated & investigated a ring-shaped alumina/PZT vibrator to form a traveling-wave USM. The rotation speed of the alumina/PZT motor was larger than that of the stainless-steel/PZT motor, meanwhile, it exhibited superior maximal-torque-to-voltage and maximal-output-power-to voltage ratios [102].

Stable operation is one of the most crucial requirements for resonators in vibratory gyroscopes and ultrasonic motors, but eigenvalue splitting can deteriorate operation stability. Wang et al., proposed the estimation and elimination of eigenvalue splitting and vibration instability of resonators arranged in a fashion of ring-shaped periodic structures [103]. To simplify the driving power of the inertia drive USM, a low-frequency USM driven by a 50 Hz sine wave was proposed. Wang et al., proposed a standing-wave trapezoidal ultrasonic linear motor, which consisted of a trapezoidal piezoceramic plate with slanted sidewalls and a clip fastener to achieve bidirectional linear motion [105]. A trapezoidal piezoceramic plate sized 22 x 8 x 1.5 mm^3 and provided a travel distance of 10.10 mm and an output force of 12.151 g at a driving voltage of 10 V was useful for compact products.

Patents were filed on by various inventors. For instance, 1) YANG et al., files a patent which described a multi-spoke-type ultrasonic motor, to increase output performance of the ultrasonic motor, prolong service life, and reduce manufacturing costs; [106] 2) Rosenkranz et al., for U -shaped piezo motors; [107] 3) YANG et al., for ultrasonic linear actuation device includes a mover and a plurality of stator sets; [108].

Shi et al., proposed a deep-sea linear ultrasonic motor, which took in-plane expansion mode as the working mode [109]. The influences of static seal and the pressures of water on the performance of the ultrasonic motor were studied. Performance of prototype whose velocity was measured at 214 mm/s while the water pressure was 8 MPa and the voltage signal with a frequency of 72 kHz and a voltage magnitude of 200 V. Nakajima et al., proposed a MDOF-USM consisting of a spherical stator and a rotor of various shapes [110]. Chen et al., presented a hollow, linear, nut-type USM based on two degenerate, 3rd-order bending modes in the section plane of cylinders [111]. The motor with four PZT plates reached an upward speed of 0.95 mm/s when the load force was 3 g, and the maximum thrust force was 0.35 N.

3.1.6 Improvements on linear USMs

Linear USMs are one of the most commonly used USMs among all because of the less complex design & effective driving method. Zheng et al. proposed a novel single-phase standing wave linear ultrasonic motor, which was made of a single PZT ceramic square plate with a circular hole in the center [60]. The driving mechanism of the motor was based on the combining the in plane expanding and bending modes to generate bidirectional linear motion [60]. They also proposed a miniature, ring-shaped, linear piezoelectric ultrasonic motor based on multimodal coupling operating in a single, in-plane mode [61]. This motor can produce a maximum driving force of 2.7 N, a no-load moving speed of 56 mm/s, and a high positioning resolution of 0.1 μm in open-loop control. It had advantages of simple structure, controllable micrometer-scale displacement, and large bidirectional working stroke indicated that the proposed linear motor had great potential for industrial applications for precise actuations [61]. Further a novel ring-shaped linear ultrasonic motor operating in orthogonal mode was proposed by them [62]. The motor was fabricated using a self-made high-performance PSN-PMS-PZT ceramic with the optimal composition, which had a high vibration velocity of 0.86 m/s. It exhibited a faster moving speed of 248 mm/s, a relatively large driving force of 2.6 N, and a high positioning precision of 0.2 m in open-loop control, indicating that the proposed linear motor based on self-made PSN-PMS-PZT ceramic had a great potential application for precise actuations [62].

Bai et al., proposed a two-way self-moving linear USM, which composed of a diamond-shaped metal elastic body, a piezoelectric ceramic piece and a parallel guide rail [63]. By exciting the piezoelectric ceramic sheets on both sides of the elastic body, the first order bending vibration mode was excited to realize the bidirectional movement of the motor. Under the excitation of 200Vpp, the forward and reverse frequency of the ultrasonic motor is 18.18KHz and 18.07KHz, and the forward and reverse no-load speed was 43.76 mm/s and 43.14 mm/s, respectfully. Takemura et al., developed a prototype of linear ultrasonic motor with an embedded preload mechanism [64]. The motor was driven bidirectionally by selective excitation of the second and third resonant vibration modes of the stator. The maximum velocity, thrust and power of the motor are 62.5 mm/s, 0.12 N and 1.01 mW respectively [63].

3.1.7 Energy harvesting type USMs

Wang et al., proposed an energy harvesting type ultrasonic motor in which two PZT rings were adopted in the new motor, one was bonded on the bottom surface of the stator metal body to generate the traveling wave in the stator, and the other one was bonded on the outside top surface of the stator metal body to harvest and convert into the vibration-induced energy of the stator into electric energy [83]. They further developed a novel multifunctional composite device by using one single PZT ring, in which a piezoelectric actuator, a sensor and an energy harvester are embedded [84]. The piezoelectric ceramic ring was polarized into three regions to produce the actuating, sensing and energy harvesting functions.

3.1.8 Comprehensive approaches

Liu's research group made a remarkable contribution in the development of novel ultrasonic motors in last 5 years [21–32]. their research manly focused on bonded type structure, USM with nanometer resolutions, symmetric & asymmetric structure, multi degree of freedom motors & hybrid excitation. Some of his work from year 2015 to 2020 are described below: 1) A cylindrical traveling wave ultrasonic motor using bonded-type composite beam was proposed, a new exciting mode for L-B (longitudinal-bending) hybrid vibrations using bonded-type was adopted, which requires only two pieces of PZT ceramic plates and a single metal beam; [21] 2) A crossbeam ultrasonic motor with miniature size was developed, which used a bonded PZT ceramics to excite two first bending vibration modes that are orthogonal in space. The symmetrical crossbeam assured that two vibrations have the same resonance frequency, which solved the problem of mode frequencies degeneracy; [22] 3) A new-type linear ultrasonic motor which combined two orthogonal bending vibration modes & eight pieces of PZT ceramic plates and a metal beam that includes two cone-shaped horns and a cylindrical driving foot was developed & the maximal velocity of the achieved by this motor was 735 mm/s and the maximal thrust 1.1 N; [23] 4) A ultrasonic motors having three degree of freedom using four piezoelectric ceramic plates in bonded-type structure was proposed. It took advantage of a longitudinal mode and two bending modes, different hybrids of which can realize three-DOF actuation [24]. Because of symmetric structure the resonance frequencies of the two bending modes were identical; 5) A novel single-mode linear piezoelectric ultrasonic motor based on asymmetric structure was proposed [25]. The motor adopts the combination of the first longitudinal vibration and the asymmetric mechanical structure to produce the oblique movement on the driving foot which resulted in linear output motion under the friction coupling between the driving foot and the runner; 6) A two-degrees-of-freedom ultrasonic motor, which could generate linear motions with two DOF by using only one longitudinal–bending hybrid sandwich transducer, was proposed [26]. The results indicate that the maximum no-load velocities of the motor in horizontal and vertical directions are 572 and 543 mm/s under the preload of 100 N and the voltage of 300Vp − p respectively. The maximum output forces in horizontal and vertical directions are 24 and 22 N when the preload was 200 N; 7) A cantilever ultrasonic motor with nanometer resolution was designed, fabricated and tested, & it achieved an output speed of 344.35 mm/s when the frequency and voltage were 22.7 kHz and 200 $V_{p\text{-}p}$ respectively [27]. The maximum output force was 8 N under the voltage and preload of 100 $V_{p\text{-}p}$ and 50 N & high displacement resolution of 48 nm under the resonant working state was achieved; [8] a novel spherical stator multi-DOF ultrasonic motor using in-plane non-axisymmetric mode was proposed [28]. The mechanical output characteristics around X, Y and Z axes were measured under different excitations, pre-tightening forces and loading conditions. The no-load rotary velocities of the prototype were 200 r/min, 198 r/min and 250 r/min and the maximum load torques were 10.8 Nm, 11.0 Nm and 12.3 Nm around X, Y and Z axes, respectively was achicved; [28] 9) a novel rotary stack having advantages of high precision, high stiffness, high dynamic range, and simple structure, which was not only suitable for generating the precise rotary motion, but also for exciting high-frequency rotary vibration was proposed this was required in the applications of micro–nano manipulations; [29] 10) A novel bending-bending piezoelectric actuator driven by a single-phase signal was proposed, in which the two-dimensional 8-shaped trajectory of the driving tip moved the runner [30]. This prototype could rotate a pulley (22 mm in diameter) at the maximum speed of 1373 rpm forward and 1350 rpm backward under a preload of 9 N, respectively; 11) A new method to reduce the volume of the traveling wave USM with ring-shape stator and improve its output speed, torque density, efficiency, and power density [31]. The USM obtained an output speed of 53.86 rpm under a preload of 0.69 N when the frequency and voltage were 24.86 kHz and 250 $V_{p\text{-}p}$, the maximum stall torque was tested as about 0.11 Nm. under the preload of 3.14 N. 12) A sandwich-type multi-degree-of-freedom (MDOF) ultrasonic motor with hybrid excitation was proposed [32]. The prototype achieves no-load speeds of 109.8 r/min, 107.9 r/min, and 290.8 r/min in the YOZ, XOZ, and XOY driving modes, respectively. The proposed motor employs only four pieces of lead zirconate titanate ceramics to achieve the MDOF rotations of a spherical rotor.

USMs are known for their precise motion & position control. Fen, et al., developed a novel integral terminal sliding-mode-based adaptive integral backstepping control (ITSMAIBC) to accommodate the impacts of inherent friction, hysteresis nonlinearity, model uncertainties and retain high tracking precision [73]. Chen et al., presented a new butterfly-shaped linear piezoelectric motor for linear motion [74]. In the closed loop condition the positioning accuracy of plus or minus <0.5 μm was experimentally obtained for the stage propelled by the piezoelectric motor. Xu et al., presented a novel standing wave ultrasonic stepping motor operated in radial vibration mode [75]. Metal blades of the stator and grooves of the rotor were designed for precise positioning. In order to improve the torque, the rotor was pushed by blades of the stator directly without any friction material.

Sarhan et al., proposed a tubular USM operating in single phase, with rectangular plate having in plane out of plane vibration [79]. The maximum speed and torque of the tubular USM motor was 59 rpm and 0.28 mNm at 80 V_{pp} of applied voltage. It can be used where accurate control and high resolution at low speed is required. [They further proposed a motor working with coupled in-plane and out-of-plane vibration modes of rectangular plate provided a large contact area between stator and rotor of motor which can reduce wear and enhance motor lifetime [80]. Overall dimension of prototype was 49x14 x2 mm, working frequency of motor was 49.6 kHz, no-load speed and stall force of motor are 122 rpm and 0.32 mN m at 50 V, respectively.

Mustafa et al., proposed extremum seeking control (ESC) as an adaptive seeking technique with fast convergence and high robustness to optimize the USM performance by tracking maximum efficiency states [243]. The application of a non-sinusoidal periodic excitation voltage to induce a near-square-wave driving tip trajectory in linear ultrasonic motors (LUSMs) was proposed by Le et al. [244]. This would reduce lost power in the periodic driving tip motion, thereby, increasing the output force and power of the LUSM. A high-efficiency Pseudo-Full-Bridge inverter with the aid of the soft-switching technology, which was accomplished by the resonance of the in-series inductance with the snubber capacitance was presented by shi et al. [245]. The efficiency of the whole drive increases by a factor of 1.25 after replacing the traditional inverter with the proposed one. A method for adjusting difference between the longitudinal and bending mode frequencies of the laminated composite stator was proposed by Li et al. [246]. The frequency adjustment method was realized by changing the applied magnetic field which affected the effective elastic modulus of the composite stator. The sensitivities of motor performances on the pre-pressure were analyzed and a targeted optimization method was discussed by Chen et al. [247]. A simulation model with power dissipation and an integrated experimental facility with the preload adjustment device was adopted to analyze the laws from multiple perspectives. Peng et al., presented a new kind of the rotary ultrasonic motor with a longitudinal vibration model of the Langevin transducer acting as the stator, while the rotor consisted of a shaft and spiral fins [248]. A high-efficiency compensation method of the dead zone with the aid of the adaptive dither for the ultrasonic motor was proposed by Shi et al. [249]. The method not only could effectively compensate the dead zone and conveniently control the velocity, but also superior to the existing phase-difference-based method in terms of improving the efficiency of the ultrasonic motor. Zhu, et al., presented a novel linear piezoelectric motor suitable for rapid ultra-precision positioning [251]. By changing the input signal, the motor could work in the fast-driving mode as well as in the precision positioning mode. In the fast-driving mode, the motor achieved maximum no-load speed of 181.2 mm/s and maximum thrust of 1.7 N at 200 V_{p-p}. & in precision positioning mode, the motor acted as a flexible hinge piezoelectric actuator producing motion in the range of 8 μm. Li et al., proposed a novel dual-frequency asymmetric excitation method for motors which can operate under traditional single-phase asymmetric or two-phase symmetric excitation modes [252]. The motor demonstrated acceptable temperature characteristics and operating stability under the proposed excitation method with calculated optimal frequencies. Zhou et al., presented a novel linear ultrasonic motor with two operation modes wherein the stator of the motor was divided into two transducers and two isosceles triangular beams [253]. Dong et al., proposed a new equivalent circuit of a piezoelectric ring in radial vibration mode considering three types of fundamental losses, i.e., dielectric, elastic, and piezoelectric. Prototype achieved a maximum torque of 270 Nm [254].

3.2 Piezoelectric materials in USM

Piezoelectric materials used in USM, plays an important role in determining its performance. In the following table we tabulated some prominently used material & their properties (**Table 3**).

Wu et al. explored how elliptical shapes and force factors of the polymer-based vibrators vary as several key structural parameters were changed [229]. Subsequently, attempt to improve the maximum torques of the polymer-based USMs by adjusting several key dimensions, and the reason for their relatively low output torques and power compared to the metal-based USMs were done. Further he employed a high-order bending mode in the polymer-based cylindrical ultrasonic motor, because this mode yields a relatively high electromechanical coupling factor, which may lead to high output power of the motor. Additionally, in contrast with the low-order modes with only vertical nodal lines, the high-order mode has both horizontal and vertical nodal lines on the circumferential outer surface of the polymer-based vibrator [230]. Similarly Wu et al. also researched on, polyphenylene sulfide (PPS)-based bimodal piezoelectric motor. Considering the viscoelasticity of PPS, the electromechanical coupling analytical model was established to describe the dynamics of the PPS-based motor by using the Kelvin-Voigt viscoelastic model. Based on the proposed model, the Taguchi method was adopted to match the resonance frequencies of the longitudinal and bending vibration. The performance test demonstrates that the PPS-based motor could yield the maximal torque of 2 mNm with the stator weight of 5.4 g [232]. Wang et al., fabricated a lead zirconate titanate (PZT) thick-film piezoelectric micro stator based on a

Sr. No.	Material	Properties	Description	Ref.
1	Poly(phenylene sulfide) (PPS)	low density, low elastic modulus, & low mechanical loss	It was used to fabricate an annular elastomer with teeth and was glued a piece of piezoelectric-ceramic annular disk to the bottom of the elastomer to form a vibrator.	[229]
2	Poly phenylene sulfide/alumina/PZT triple-layered	low density, low elastic modulus, & low mechanical loss	A thin alumina disk sandwiched between the PPS vibrating body and PZT disk to compensate the stiffness was constructed. The maximum output torque and power of the triple-layered motors were 5 and 13 times the values of the double-layered motors, respectively, due to the enhanced force factor and electromechanical coupling factor.	[232]
3	Lead-free CH doped (K0.5 Na0.5)NbO3 (KNN) ceramics	kp:34.1% (±2%); kt:45.3% (±2%); Qm:3170 (±2%); Rz:8.6 Ω (± 3%); and tanδ:0.1%.	Piezoelectric motors fabricated using these ceramics achieved a velocity of 4.5 mm/s, vertical velocity of 3.02 mm/s, and output power of 2.93 mW with a negligible increase in temperature and high stability while driven.	[234]
4	c-axis crystal-oriented (Sr,Ca)2NaNb5O15 (SCNN) plate	mechanical coupling coefficient of k31 = 7%, a mechanical quality factor of Qm = 3200	The mechanical nonlinearity behavior was not observed, and the temperature dependences of the quality factor and the equivalent stiffness decreased. Motor shown the revolution speed 100 rpm in torque from 150 μN·m to 900 μN·m, the output power 7.5 mW, efficiency 3.5% at 657 μN·m and 109 rpm.	[240]
5	polyimide (PI) composite (1.41 g/cm3) reinforced with carbon fibers (CF)	high elastic modulus, wear resistance, and suitable friction coefficient.	To reduce the weight without decreasing the mechanical output performance this material was used for making Stator. Output Stall Torque = 0.22 Nm, Wear Resistance 1.38×10^{-5} mm/N·m, reduce the weight over 83.6%	[237]
6	Mn doped 0.27PIN-0.46PMN-0.27PT single crystal	lower excitation frequency, lower driving voltage, and less power loss. Mechanical quality factors (Qm ~ 600)	USM using this material in single-mode was made. Motor speed up to 42.3 cm/s, a driving torque of 0.42 N cm, and an output power density of 0.45 W/cm3, under the driving voltage of 21 V was obtained. Used for making miniaturized and high-power motors.	[238]

Sr. No.	Material	Properties	Description	Ref.
7	single crystal lead magnoniobate titanate (PMNT)	excellent piezoelectric properties and temperature stability	The criterions of crystal orientation for ring type USM proposed. Cutting orientation for the crystal poled along $[001]_c$ direction has better electromechanical properties and process compatibility. This orientation improves the lateral piezoelectric coefficient from ~90 pC/N to ~1201 pC/N and electromechanical coupling factor to 0.92.	[239]
8	Surface textured polyimide composites	High friction coefficient and high elastic modulus	To enhance conversion efficiency & tribological performance. The surface texture is capable of storing abrasive debris to protect the friction interface. Meanwhile, boundary effect of texture can increase friction coefficient. After the combination of the advanced PI friction material and the surface texture, the conversion efficiency of the USM increased by 82.8%.	[241]

Table 3.
The list of piezoelectric materials for USM.

high-performance PZT thick film by electrohydrodynamic jet (E-jet) printing in order to simplify the manufacturing process and enhanced the performance of the piezoelectric stator. The thick-film micro stator produced a traveling wave with an amplitude of 345 nm, and the mechanical quality factor was found to be 736 [236]. Zhao et al., proposed an oblate-type ultrasonic micro-motor with multilayer piezoelectric ceramic with chamfered driving tips. The micro-motor works based on the standing-wave principle and has a higher rotary speed than the traditional standing-wave. The experimental results showed that the rotary speed was around 2000 r/min at the voltage of 20 V_{p-p} [242].

3.3 Thermal performance of USM

Performance of Ultrasonic motors in extreme temperature setting is the key challenge faced by various researchers. Thus, Nishizawa et al. developed spherical ultrasonic motor for space application & investigated for the durability to the radiant heat from the sun [256]. Drive performance was conducted for estimated duration more than 70 mins for higher than +120°C conditions. In order to maintain its drive performance, selection of piezoelectric elements & adhesive materials were significantly discussed. [256], further spherical USM was investigated in low temperature environment of −80°C, & approximately 60 minutes cumulative drive time was achieved by applying the same piezoelectric element and the adhesive materials utilized for high temperature conditions [257]. Shi et al., presented a general optimum frequency tracking scheme for an ultrasonic motor, which no longer required the amplitudes of the applied voltages to keep identical [258]. The mechanical quality factor of an ultrasonic motor was initially derived to describe

Applications	Description
Magnetic Resonance Imaging	To determine the effects of a USM on MR images in the high field MRI scanner (3 T) using signal-to-noise ratio (SNR) [268].
Magnetic Resonance Imaging	To evaluate the temperature increase caused by a 3.0-T magnetic resonance imaging (MRI) system on an ultrasonic motor (USM) used to actuate surgical robots in the MRI environment [269].
Magnetic Resonance Imaging	To quantify and compensate the geometric distortion of MR images as generated by the presence of USMs [270]
Magnetic Resonance Imaging	To investigate displacement force and torque applied to an ultrasonic motor at various bore locations using the designed apparatus presented in the Materials and Methods [271]
Magnetic Resonance Imaging	to study the types of image artifacts generated by the USM, provide comparison between them, introduce their sources, and provide compensation methods [272].
Magnetic Resonance Imaging	To demonstrate that image distortion related issues can be partly addressed by replacing metallic nonactive motor components from a resonant ultrasonic motor for non-metallic equivalents [273].
Magnetic Resonance Imaging	A new cable-driven robot for MRI-guided breast biopsy. A compact three degree-of-freedom (DOF) semi-automated robot driven by ultrasonic motors was designed with non-magnetic materials [274].
Wearable Walking assist device	The fundamental study for understanding and quantifying the muscle fatigue reduction effect of walking assist system which is applicable to individual characteristic using ultrasonic motors [275].
Wearable Walking assist device	A hip-joint support ambient walking assistive system consider a timing and amplitude of assist for effective and individual-oriented assist was investigated [276].
Medical Endoscope	A spiral motion hollow micromotor operating in E02-mode traveling wave with outer diameter (3.6 mm) and length (3 mm), applied to an endoscope imaging system for driving optical lens, and a high quality image has been obtained due to its autofocus & autozoom function [277].
Low-frequency Sonophoresis system of transdermal drug delivery.	A new sonophoresis system of transdermal drug delivery, which involves the combination of an ultrasonic transducer and a linear ultrasonic motor, is designed to control permeability in in vitro LFS (Low frequency sonophoresis) [278]
Surgical Instrument with Torque Assist	To an ultrasonic surgical instrument including a torque assist feature to facilitate connection of the waveguide with an ultrasonic transducer [279].
Humanoid eyeball orientation system	A compact ring type 3-DOF USM fabricated to meet the specification of requirements for humanoid eyeball orientation system [280].
Underwater Robots	A dual-rotor ultrasonic motor with double output shafts, compact size, and no electromagnetic interference, characterized, and applied for actuating underwater robots [281].
Deep Sea Drones	A prototype of deep-sea drone by use of spherical ultrasonic motors for sensing marine bottom and make maps of the 4,000 m depth grades [282].
Single-gimbal control moment gyroscope	A structure-compacted single-gimbal control moment gyroscope, directly driven by an ultrasonic motor obtained wide-range closed-loop speed control of the gimbal from 0.2 mrad/s to 1.6 rad/s, which is the base of the high stability and high precision of the control moment gyroscope [283].
Continuously variable beam expander	An automatic continuously variable beam expander with two hollow ultrasonic motors as its actuators was used to expand a laser beam by between threefold and fivefold, and nanoscale positioning and high-precision beam shaping was achieved [284].
Two-axis Nonmagnetic Turntable	Non-magnetic technology to carry out research on the pointing accuracy and motion control of the turntable [285].

Applications	Description
Morphing carbon fiber composite airfoil	A morphing carbon fiber composite airfoil concept with an active trailing edge enabled by an innovative structure driven by an electrical actuation system that uses linear ultrasonic motors (LUSM) with compliant runners, providing full control of multiple degrees of freedom [286].
Bio-Inspired Flapping Wing Rotor	To vary the flapping frequency rapidly during a stroke, an ultrasonic motor (USM) was used to drive the FWR.(Flapping Wing Rotor) [287].
Blade Free Drone	A "Blade-Free drone" is a blimp style drone for safe indoor flying without the use of propellers or flapping wings. It uses micro blowers which can eject air through ultrasonic vibrations generated by piezoelectric device as an actuator [288].
LiDAR Systems	A novel design of a compact standing-wave rotary USM based on a coupled axial-tangential mode, one power source, and a bulk piezoelectric actuator of a tube form was developed for autonomous vehicle [289].
An Optical Image Stabilization of Portable Digital Camcorders	A sensor-shift optical image stabilization (OIS) using a novel ultrasonic linear motors (ULMs) and a fuzzy sliding-mode controller (FSMC) was used to compensates the optical path deviation that was caused by user's hand-tremor for avoiding image blurring [290].
Dual piezoelectric beam robot	To study the effect of the piezoelectric patches' positions on the performance of the robot [291]
Haptic grippers	Two piezometers, walking and traveling-Wave Piezoelectric Motors used as Actuator in haptic grippers were compared. Walking quasi-static motor was superior at low velocities & traveling wave ultrasonic motors were suitable when high velocity is required [292].
Microgripper	3-DOF microgripper driven by Linear Ultrasonic Motors (LUMs), with nanometer positional accuracy and an operational space in the millimeter scale was presented [293].
Variable Aperture	A new type of piezoelectric actuator with a screw-coupled stator and rotor was developed to operate an aperture. The actuator and the aperture are integrated to control the luminous flux. This actuator is having high resolution, high speed, simple structure and compact size [294].
Coiled stator ultrasound motor (CS-USM)	The propagation velocity of elastic waves from the simulated the vibration displacement mode profile along a straight line acoustic waveguide was analyzed via three-dimensional finite element method [295]
Scanning probe microscopy	One coordinate piezoelectric stepping motor of the scanning probe microscopy (SPM) nanomanipulator, allocated to the positioning was designed [296].
Gravimeter	A novel absolute gravimeter based on an ultrasonic motor was proposed in order to resolve the limitations related to free-fall absolute gravimeter, such as the complex structure and its large site [297].
photoacoustic microscopy	A PAM (photoacoustic microscopy) system was designed with a miniature ultrasonic actuator to significantly downsize the scanning probe (61 mm × 50 mm × 47 mm) [298]

Table 4.
Listed of the research articles (2015–2020) on the applications of USM.

the loss, which further was also in proportion to the temperature rise. The optimum frequency from the loss reduction viewpoint was then obtained, at which frequency the ultrasonic motor maintained the minimum loss and subsequently the minimum temperature rise. Sunif et al., presented heat energy modeling method for determining and characterizing of a piezoelectric stator profile that applied in a piezoceramic ultrasonic motor with the consideration of heat generated [259]. A thermal analysis was conducted in order to analyze the heat distribution on the stator & results showed different longitudinal deflection with the increment of the temperature. Liu et al. studied the temperature variations of different components under different driving voltages for a high-power longitudinal-longitudinal hybrid type T shaped ultrasonic motor [260]. Cheng et al. described about hypothesis that a temperature gradient transverse to the wave propagating direction could significantly increase the working velocity of acoustic streaming-driven motors which was then investigated by numerically solving the hydrodynamic equations & it was found that the velocity of the rotor only weakly depends on the transverse temperature gradient, the velocity increased by only ~8.8% for temperature difference of 40°C between the rotor and the stator [261]. Nakazono et al., studied temperature dependence of USM in cryogenic conditions [262]. Ultrasonic transducer comprising of a body & nut made of SUS304 & bolt made of titanium was fabricated & evaluated in the temperature range of 45 to 293 K. It was proved that when titanium was used for clamping bolt of the transducer, the motor can be driven without the regulation of the preload.] Lv et al., developed a novel theoretical model to investigate the temperature field and output characteristics of a standing wave ultrasonic motor [263]. The results showed that the developed model can not only predict the temperature variation of motor in continuous operation but also evaluate the influence of surface roughness and various input parameters on output characteristics of motor.

We reviewed some literature review papers published on ultrasonic motors. We found author Peng et al. reviewed literature & provided summary on precision piezoelectric motors over long ranges based on the principle of repeating a series of small periodic step motions, named "frequency leveraged motors" [264]. Work was classified into three categories by different frequency driving methods, including ultrasonic motors, quasi-static motors, and motors combined resonant and quasistatic operations. A comprehensive summary of piezoelectric motors, with their classification from initial idea to recent progress, was presented by Spanner and Koc [265]. This review also includes some of the industrial and commercial applications of piezoelectric motors that are presently available in the market as actuators. Peled et al., reviewed & provided summary of the design of high precision motion solutions based on L1B2 (first longitudinal and second bending modes) ultrasonic motors—from the basic motor structure to the complete motion solution architecture, including motor drive and control, material considerations and performance envelope [266]. Gao et al., presented recent progress in non-resonance piezoelectric actuators with the working principles and properties of actuators and the piezoelectric materials and configurations, fabrication, and applications [267].

3.4 Ultrasonic motors applications

The **Table 4** illustrates about various research articles published on the ultrasonic motor applications.

4. Industry 4.0

Kagermann in 2011 first published the main ideas of Industry 4.0 [299] and built the foundation for the Industry 4.0 manifesto published in 2013 by the German National Academy of Science and Engineering (acatech) [300, 301]. The concept of Industry 4.0 is based on the integration of information and communication technologies and industrial technology and is mainly dependent on building a Cyber-Physical System (CPS) to realize a digital and intelligent factory, in order to promote manufacturing to become more digital, information-led, customized, and green. The purpose of Industry 4.0 is to build a highly flexible production model of personalized and digital products and services, with real-time interactions between people, products and devices during the production process [302]. Industry 4.0 is a complex and flexible system involving digital manufacturing technology, network communication technology, computer technology, automation technology and many other areas [302]. There are many technologies used to implement Industry 4.0 like Internet of things, cybersecurity, augmented reality, big data & AI Analytics, Autonomous robot, additive manufacturing, Simulation & Digital twin, System integration & cloud computing [303]. Among all this we will be considering a few of them for improving the performance of the USM & we will also predict, how USM can best fit into industry 4.0 scenario.

Digital twin plays a role of a bridge between cyber world & physical world [304]. A digital twin of USM can be made to monitor its development from design phase to end user. It can be further extended up to recycling & remanufacturing & complete cyber physical system can be implemented [305]. Number of components used, CAD drawing of mechanical & electrical component, materials & their properties can be embedded in the 3D model, this will be the first step in formation of the digital twin of USM [305]. Results of simulation of the motor in various environment to validate its performance & product design can also be embedded in the digital twin to enhance it further. Additionally, the environmental performance and impact can be simulated at this phase via life cycle assessment (LCA) module too, e.g., design for recycling, design for disassembly and design for remanufacturing. The simulation results and disassembly are maintained in the product archive for future remanufacturing operations [305]. When the USM will be sold to an end user, he or she can update the product status via various Industry 4.0 enablers, e.g., mobile apps, smart tags, QR code, websites, and so forth. At this phase, the changes of product, e.g., location, ownership, upgrading, repairing and maintenance, can be updated and maintained inside the mirrored digital twin [305]. When the USM stops service, the users can update the digital status via mobile app or web service. He or she can contact a professional collector who has the expertise in this specific device. Failed USM can be evaluated, and testing can be applied to the individual component. Based on the examination results, the digital twin can be updated, and the proper operation can be planned accordingly, e.g., recovery at the component level, material level [305]. Developing a holistic digital twin of USM which captures data from the real world will be highly useful for tackling many challenges encountered during the design, modeling & optimization phase of the motor. Thus, action can be taken accordingly during the design phase in order to enhance the performance of the motor. So that, fully developed digital twin model of USM can be further utilized to simulate it for various types of application i.e., space exploration, medical devices, manufacturing industries etc [1].

Large amount of the data is generated during simulation & validation or collected from the digital twin of the USM or from the end user. This data can be used for analyzing & enhancing the performance of the USM. Further this data can be used to create algorithm which can be used for precise motion & drive control of the motors thus improving its overall performance. Depending on the application, a variety of algorithms can be used such as artificial neural networks (ANN) [306, 307]. ML offers great potential for intelligent data analyses and is a key technology for autonomous robots, image and signal analysis as well as complex controls for sensor-actuator systems [306, 308]. ML techniques can e.g., contribute to condition monitoring, predictive maintenance, or process control of the motor [306–308]. Consequently, data generated during operation of USM can be effectively utilized as important information for various sensors installed in Industry 4.0 setup [300].

Additive manufacturing can play a vital role in designing the complex shapes & size of the USM for different application. In AM components are manufactured by layer-by-layer deposition of the material [306, 309]. The possibility of materializing a complex digital model directly into a physical component without the need for shaping, maintaining and warehousing tools as well as manual intervention ideally reflects the idea of digital production [306]. AM allows the production of geometrically complex, function-optimized and customer-specific products at any location equipped with a suitable AM machine and thus contributes to the flexibilization and globalization of production processes [306]. Further, AM offers potential business for making spare parts [306, 310]. By using AM, designer can create physical prototype of the components used in USM using different combination of materials to analyze its fit, form & function for desired application.

Using IoT (internet of things) and CPSs (cyber physical system) it is possible to monitor the motors in real time. Interconnection of motors & the data obtained while its running during real time makes it possible to react quickly, effectively to every instance. Sensors embedded in the motors can give real time feedback of the parameters for instance, temperature rise which can be further analyzed for improving the design & performance of the motor. These sensors can also provide data for the whole life cycle of the motor which can be thus utilized by the researchers & manufacturers to incorporate into the motors which will ultimately result into overall improvement of the motor thus saving time & energy [300].

Augmented reality is the key technology of Industry 4.0 [311]. It enables human to access digital information and overlay that information with the physical world [312]. Ultrasonic motors are used in various engineering application areas i.e., from medical field to aerospace. They are manufactured in various size & shapes [1]. Use of augmented reality (AR) will make the USM designer aware of the fit, form & function of the motor suitable for different types of application w.r.t volume & space availability in real world environment. Thus, by using AR technology they can effectively design the USM based upon the requirement.

5. Conclusions

This book chapter gives a brief summarization of the research articles from eclectic sources like journal, patents and masters-PhD thesis published in last 5 years on piezoelectric ultrasonic motors. This article gives us the statistical analysis of number of publications on ultrasonic motors with respect to year of publication, country of origin & list of top 10 research journal. It divides all the articles into different categories and highlights the challenges, opportunities & research carried out. It broadly classifies the research article in areas like new design, modeling & simulation, friction & wear, piezoelectric materials, thermal performance & USM applications. Further it introduces with concept of Industry 4.0 & its key enabling technologies. It explains how to apply industry 4.0 technologies like digital twins & simulations, big data & machine learning, Industrial internet of things, additive manufacturing & augmented reality to improve the design & performance of the USM and how USM can best fit in Industry 4.0 era. Ultrasonic motors have tremendous potential for the improvement & Industry 4.0 technologies has tremendous potential to improve its design & performance on the other hand USM has many advantages & wide application areas which can be best suited for industry 4.0 settings. Thus, both USM & industry 4.0 if amalgamated together can give us a remarkable output in future.

Conflict of interest

NA

Notes/Thanks/Other declarations

NA

Author details

Sahil P. Wankhede and Tian-Bing Xu*
Department of Mechanical Engineering and Aerospace, Old Dominion University, Norfolk, VA, USA

*Address all correspondence to: txxu@odu.edu

IntechOpen

© 2021 The Author(s). Licensee IntechOpen. This chapter is distributed under the terms of the Creative Commons Attribution License (http://creativecommons.org/licenses/by/3.0), which permits unrestricted use, distribution, and reproduction in any medium, provided the original work is properly cited.

References

[1] Zhao C, Ultrasonic Motors Technologies and applications. Springer, 2011, DOI: https://doi.org/10.1007/978-3-642-15305-1

[2] Tian X, Liu Y, Deng J, Wang L, Chen, W. A review on piezoelectric ultrasonic motors for the past decade: Classification, operating principle, performance, and future work perspectives. Sens Actuators Phys. 2020 May; 306:111971.

[3] Mashimo T. Performance evaluation of a micro ultrasonic motor using a one-cubic-millimetre stator. IEEE Trans Ultrason. Ferroelectric. Freq Control. 2015 Oct;62(10):1819-1826. doi: 10.1109/TUFFC.2014.006834.

[4] Mashimo T. Micro Ultrasonic Motor Using a Cube with a Side Length of 0.5 mm. IEEEASME Trans. Mechatron. 2016 Apr;21(2):1189-1192. doi: 10.1109/TMECH.2015.2511782.

[5] Mashimo T. Miniature preload mechanisms for a micro ultrasonic motor. Sens Actuators Phys. 2017 Apr;257:106-112. doi: 10.1016/j.sna.2017.02.009.

[6] Mashimo T. Scaling of Piezoelectric Ultrasonic Motors at Submillimetre Range. IEEEASME Trans Mechatron. 2017 Jun;22(3):1238-1246. doi: 10.1109/TMECH.2017.2691805.

[7] Mashimo T, Urakubo T, Shimizu Y. Micro Geared Ultrasonic Motor. IEEEASME Trans Mechatron. 2018 Apr;23(2):781-787. doi: 10.1109/TMECH.2018.2792462.

[8] Izuhara S, Mashimo T. Design and evaluation of a micro linear ultrasonic motor. Sens Actuators Phys. 2018 Aug;278:60-66. doi: 10.1016/j.sna.2018.05.022

[9] Miyoshi K, Mashimo T. Miniature direct-drive two-link using a micro-flat ultrasonic motor. Adv Robot. 2018 Oct 18;32(20):1102-1110. doi: 10.1080/01691864.2018.1524313.

[10] Mashimo T, Izuhara S, Arai S, Zhang Z, Oku H. High-Speed Visual Feedback Control of Miniature Rotating Mirror System Using a Micro Ultrasonic Motor. IEEE Access. 2020;8:38546-38553. doi: 10.1109/ACCESS.2019.2957298.

[11] Hareesh P, DeVoe DL. Miniature Bulk PZT Traveling Wave Ultrasonic Motors for Low-Speed High-Torque Rotary Actuation. J Microelectromechanical Syst. 2018 Jun;27(3):547-554. doi: 10.1109/JMEMS.2018.2823980.

[12] Yang X, Liu Y, Chen W, Liu J. Miniaturization of a longitudinal–bending hybrid linear ultrasonic motor. Ceram Int. 2015 Jul;41:S607–S611. doi: 10.1016/j.ceramint.2015.03.248.

[13] Yu H, Quan Q, Tian X, Li H. Optimization and Analysis of a U-Shaped Linear Piezoelectric Ultrasonic Motor Using Longitudinal Transducers. Sensors. 2018 Mar 7;18(3):809. doi: 10.3390/s18030809.

[14] Wang F, Nishizawa U, Tanaka H, Toyama S. Development of miniature spherical ultrasonic motor using wire stators. J Vibroengineering. 2018 Dec 31;20(8):2939-50. doi: 10.21595/jve.2018.20348.

[15] Ho S-T, Chiu W-H. A piezoelectric screw-driven motor operating in shear vibration modes. J Intell. Mater Syst. Struct. 2016 Jan;27(1):134-145. doi: 10.1177/1045389X14563863.

[16] Ran L, Zhou W, He J, Zhan L, Chen Q, Yu H, et al. A novel three-dimensional contact model of piezoelectric traveling wave ultrasonic micromotor. Smart Mater Struct. 2020

Jul 1;29(7):075016. doi: 10.1088/1361-665X/ab87e3.

[17] Wu J, Mizuno Y, Nakamura K. A rotary ultrasonic motor operating in torsional/bending modes with high torque density and high-power density. IEEE Trans Ind Electron. 2020;1-1. doi: 10.1109/TIE.2020.3000112.

[18] Zhou Y, Chang J, Liao X, Feng Z. Ring-shaped traveling wave ultrasonic motor for high-output power density with suspension stator. Ultrasonics. 2020 Mar; 102:106040. doi: 10.1016/j.ultras.2019.106040.

[19] Fan P, Shu X, Yuan T, Li C. A novel high thrust–weight ratio linear ultrasonic motor driven by single–phase signal. Rev Sci Instrum. 2018 Aug;89(8):085001. doi: 10.1063/1.5037407. doi: 10.1063/1.5037407.

[20] Li H, Wang L, Cheng T, He M, Zhao H, Gao H. A High-Thrust Screw-Type Piezoelectric Ultrasonic Motor with Three-Wavelength Exciting Mode. Appl Sci. 2016 Dec 16;6(12):442. doi: 10.3390/app6120442. doi: 10.3390/app6120442

[21] Yang X, Liu Y, Chen W, Liu J. A cylindrical traveling wave ultrasonic motor using bonded-type composite beam. Ultrasonics. 2016 Feb; 65:277-281. doi: 10.1016/j.ultras.2015.09.014.

[22] Xu D, Liu Y, Liu J, Chen W. A Bonded Type Ultrasonic Motor Using the Bending of a Crossbeam. IEEE Access. 2016; 4:1109-1116. doi: 10.1109/ACCESS.2016.2542861.

[23] Yan J, Liu Y, Liu J, Xu D, Chen W. The design and experiment of a novel ultrasonic motor based on the combination of bending modes. Ultrasonics. 2016 Sep; 71:205-210. doi: 10.1016/j.ultras.2016.07.002.

[24] Yan J, Liu Y, Shi S, Chen W. A three-DOF ultrasonic motor using four piezoelectric ceramic plates in bonded-type structure. J Vibroengineering. 2018 Feb 15;20(1):358-67. doi: 10.21595/jve.2017.18523.

[25] Wang L, Liu J, Liu Y, Tian X, Yan J. A novel single-mode linear piezoelectric ultrasonic motor based on asymmetric structure. Ultrasonics. 2018 Sep; 89:137-142. doi: 10.1016/j.ultras.2018.05.010

[26] Liu Y, Yan J, Wang L, Chen W. A Two-DOF Ultrasonic Motor Using a Longitudinal–Bending Hybrid Sandwich Transducer. IEEE Trans Ind Electron. 2019 Apr;66(4):3041-3050. doi: 10.1109/TIE.2018.2847655

[27] Wang L, Guan Y, Liu Y, Deng J, Liu J. A compact cantilever type ultrasonic motor with nanometer resolution: design and performance evaluation. IEEE Trans Ind Electron. 2020;1-1. doi: 10.1109/TIE.2020.2965481.

[28] Huang Z, Shi S, Chen W, Wang L, Wu L, Liu Y. Development of a novel spherical stator multi-DOF ultrasonic motor using in-plane non-axisymmetric mode. Mech Syst Signal Process. 2020 Jun; 140:106658. doi: 10.1016/j.ymssp.2020.106658.

[29] Yu H, Liu Y, Deng J, Zhang S. A Novel Piezoelectric Stack for Rotary Motion by d_{15} Working Mode: Principle, Modeling, Simulation, and Experiments. IEEEASME Trans Mechatron. 2020 Apr;25(2):491-501. doi: 10.1109/TMECH.2020.2965962.

[30] Tian X, Liu Y, Deng J, Chen W. Single-phase drive bending-bending piezoelectric actuator operated under 8-shaped trajectory vibration: Concept, computation and experiment evaluation. Mech Syst Signal Process. 2020 May; 139:106637. doi: 10.1016/j.ymssp.2020.106637.

[31] Ma X, Liu J, Deng J, Liu Q, Liu Y. A Rotary Traveling Wave Ultrasonic Motor

With Four Groups of Nested PZT Ceramics: Design and Performance Evaluation. IEEE Trans Ultrasonic Ferroelectric Frequency Control. 2020 Jul; 67(7):1462-1469. doi: 10.1109/TUFFC.2020.2972307.

[32] Yang X, Liu Y, Chen W, Liu J. Sandwich-Type Multi-Degree-of-Freedom Ultrasonic Motor with Hybrid Excitation. IEEE Access. 2016; 4:905-913. doi: 10.1109/ACCESS.2016.2536611.

[33] Shi S, Xiong H, Liu Y, Chen W, Liu J. A ring-type multi-DOF ultrasonic motor with four feet driving consistently. Ultrasonics. 2017 Apr; 76:234-244. doi: 10.1016/j.ultras.2017.01.005.

[34] Sun M, Wang Y, Huang W, Lu Q. Research on a symmetric non-resonant piezoelectric linear motor. J Vibroengineering. 2016 Aug 15; 18(5):2916-25. doi: 10.21595/jve.2016.16599.

[35] Wang L, Wielert T, Twiefel J, Jin J, Wallaschek J. A rod type linear ultrasonic motor utilizing longitudinal traveling waves: proof of concept. Smart Mater Struct. 2017 Aug 1;26(8):085013. doi: 10.1088/1361-665X/aa78d2

[36] Sun B-J. 1643. Structure design of linear ultrasonic motor with a laminated U-shaped stator based on modal analysis. ISSN. 2015;17(4):12.

[37] Li X, Yao Z, Yang M. A novel large thrust-weight ratio V-shaped linear ultrasonic motor with a flexible joint. Rev Sci Instrum. 2017 Jun;88(6):065003. doi: 10.1063/1.4985703.

[38] Zhang B, Yao Z, Liu Z, Li X. A novel L-shaped linear ultrasonic motor operating in a single resonance mode. Rev Sci Instrum. 2018 Jan;89(1):015006. doi: 10.1063/1.5011427.

[39] Jian Y, Yao Z, Zhang B, Liu Z. A novel Π-type linear ultrasonic motor driven by a single mode. Rev Sci Instrum. 2018 Dec;89(12):125010. doi: 10.1063/1.5055278.

[40] Wang Y, Wang L. Design, fabricate, and experimental verification of an ultrasonic linear motor derived from V-type motors. Rev Sci Instrum. 2020 Apr 1;91(4):045002. doi: 10.1063/1.5129586.

[41] Li Z, Guo P, Wang Z, Zhao L, Wang Q. Design and Analysis of Electromagnetic-Piezoelectric Hybrid Driven Three-Degree-of-Freedom Motor. Sensors. 2020 Mar 14;20(6):1621. doi: 10.3390/s20061621.

[42] Li Z, Wang Z, Guo P, Zhao L, Wang Q. A ball-type multi-DOF ultrasonic motor with three embedded traveling wave stators. Sens Actuators Phys. 2020 Oct; 313:112161. doi: 10.1016/j.sna.2020.112161.

[43] Li, Z., Zhao, L., Wang, Z. et al. Traveling Wave Type Multi-Degree-of-Freedom Spherical Ultrasonic Motor with Built-in Stators. J. Electr. Eng. Technol. 15, 1723-1733 (2020). https://doi.org/10.1007/s42835-020-00463-0

[44] Jūrėnas V, Kazokaitis G, Mažeika D. 3DOF Ultrasonic Motor with Two Piezoelectric Rings. Sensors. 2020 Feb 4;20(3):834. doi: 10.3390/s20030834.

[45] Niu R, Liu J, Zhu H, Zhao C. Design and evaluation of a novel light arc-shaped ultrasonic motor. AIP Adv. 2019 Jun;9(6):065009. doi: 10.1063/1.5100794.

[46] Liu J, Niu Z-J, Zhu H, Zhao C-S. Design and Experiment of a Large-Aperture Hollow Traveling Wave Ultrasonic Motor with Low Speed and High Torque. Appl Sci. 2019 Sep 23;9(19):3979. doi: 10.3390/app9193979.

[47] Liu J, Niu R, Zhu H, Zhao C. Improving the efficiency of a hollow ultrasonic motor by optimizing the stator's effective electromechanical

coupling coefficient. Rev Sci Instrum. 2020 Jan 1;91(1):016104. doi: 10.1063/1.5121840.

[48] Niu R, Zhu H, Zhao C. A four-legged linear ultrasonic motor: Design and experiments. Rev Sci Instrum. 2020 Jul 1;91(7):076107. doi: 10.1063/1.5114787

[49] Izuhara S, Mashimo T. Linear piezoelectric motor using a hollow rectangular stator. Sens Actuators Phys. 2020 Jul; 309:112002. doi: 10.1016/j.sna.2020.112002.

[50] Zhang J, Yang L, Ma C, Ren W, Zhao C, Wang F. Improving efficiency of traveling wave rotary ultrasonic motor by optimizing stator. Rev Sci Instrum. 2019 May;90(5):056104. doi: 10.1063/1.5088804.

[51] Yang L, Ren W, Ma C, Chen L. Mechanical simulation and contact analysis of the hybrid longitudinal-torsional ultrasonic motor. Ultrasonics. 2020 Jan;100:105982. doi: 10.1016/j.ultras.2019.105982.

[52] Ren W, Yang M, Chen L, Ma C, Yang L. Mechanical optimization of a novel hollow traveling wave rotary ultrasonic motor. J Intell Mater Syst Struct. 2020 May;31(8):1091-1100. doi: 10.1177/1045389X20910263.

[53] Jian L, Wang J, Chen C. Analysis of the characteristics of traveling wave ultrasonic motor considering the structural stiffness of the preload structure. AIP Adv. 2019 Dec 1;9(12):125336. doi: 10.1063/1.5133421.

[54] Dong Z, Yang M, Chen Z, Xu L, Meng F, Ou W. Design and performance analysis of a rotary traveling wave ultrasonic motor with double vibrators. Ultrasonics. 2016 Sep; 71:134-141. doi: 10.1016/j.ultras.2016.06.004.

[55] Pan S, Xu Z, Chen L, Huang W, Wu J. Coupled Dynamic Modeling and Analysis of the Single Gimbal Control Moment Gyroscope Driven by Ultrasonic Motor. IEEE Access. 2020;1-1. doi: 10.1109/ACCESS.2020.3012694.

[56] An D, Yang M, Zhuang X, Yang T, Meng F, Dong Z. Dual traveling wave rotary ultrasonic motor with single active vibrator. Appl Phys Lett. 2017 Apr 3;110(14):143507. doi: 10.1063/1.4979699.

[57] Dong Z, Yang M. Optimal design of a double-vibrator ultrasonic motor using combination method of finite element method, sensitivity analysis and adaptive genetic algorithm. Sens Actuators Phys. 2017 Oct; 266:1-8. doi: 10.1016/j.sna.2017.09.006.

[58] Li C, Lu C, Ma Y, Li S, Huang W. Design of an ultrasonic motor with multi-vibrators. J Zhejiang Univ-Sci A. 2016 Sep;17(9):724-732. doi: 10.1631/jzus.A1500316.

[59] Mohammed UJ, Zakariyya RS. Research on a Linear Ultrasonic Motor with Double Cantilever Vibrators. In: 2020 IEEE 4th Information Technology, Networking, Electronic and Automation Control Conference (ITNEC) [Internet]. Chongqing, China: IEEE; 2020 [cited 2020 Aug 19]. p. 2295-8. Available from: https://ieeexplore.ieee.org/document/9085004/doi: 10.1109/ITNEC48623.2020.9085004

[60] Chen Z, Li X, Ci P, Liu G, Dong S. A standing wave linear ultrasonic motor operating in in-plane expanding and bending modes. Rev Sci Instrum. 2015 Mar;86(3):035002. doi: 10.1063/1.4914843

[61] Xin X, Gao X, Wu J, Li Z, Chu Z, Dong S. A ring-shaped, linear piezoelectric ultrasonic motor operating in E_{01} mode. Appl Phys Lett. 2020 Apr 13;116(15):152902. doi: 10.1063/5.0006524.

[62] Xin X, Yu Y, Wu J, Gao X, Li Z, Yi X, et al. A ring-shaped linear ultrasonic

motor based on PSN-PMS-PZT ceramic. Sens Actuators Phys. 2020 Jul; 309:112036. doi: 10.1016/j.sna.2020.112036.

[63] Bai K, Li C, Xi C. Design of two-way self-moving linear ultrasonic motor. IOP Conf Ser Earth Environ Sci. 2020 Mar 21; 446:042065. doi: 10.1088/1755-1315/446/4/042065.

[64] Kazumi T, Kurashina Y, Takemura K. Ultrasonic motor with embedded preload mechanism. Sens Actuators Phys. 2019 Apr; 289:44-49. doi: 10.1016/j.sna.2019.02.010.

[65] Zhou X, Chen W, Liu J. A novel multi-mode differential ultrasonic motor based on variable mode excitation. Sens Actuators Phys. 2015 Jul; 230:117-125. doi: 10.1016/j.sna.2015.04.020.

[66] Zhou X, Chen W, Liu J. Novel 2-DOF Planar Ultrasonic Motor With Characteristic of Variable Mode Excitation. IEEE Trans Ind Electron. 2016 Nov;63(11):6941-6948. doi: 10.1109/TIE.2016.2586018.

[67] Zhou XY, Chen WS, Liu JK. A new ultrasonic motor utilizing longitudinal vibration of sandwich type transducers. In: Fifth Asia International Symposium on Mechatronics (AISM 2015) [Internet]. Guilin, China: Institution of Engineering and Technology; 2015 [cited 2020 Aug 19]. p. 6 .-6. Available from: https://digital-library.theiet.org/content/conferences/10.1049/cp.2015.1565

[68] Zhou X, Zhang Y. A New Linear Ultrasonic Motor Using Hybrid Longitudinal Vibration Mode. IEEE Access. 2016; 4:10158-10165. doi: 10.1109/ACCESS.2017.2647972.

[69] Zhou X, Chen W, Liu J. A new rotary ultrasonic motor using longitudinal vibration transducers. Adv Mech Eng. 2015 May 19;7(5): 168781401558742. doi: 10.1016/j.ultras.2020.106158.

[70] Lu D, Lin Q, Chen B, Jiang C, Hu X. A single-modal linear ultrasonic motor based on multi vibration modes of PZT ceramics. Ultrasonics. 2020 Sep; 107:106158. doi: 10.1016/j.ultras.2020.106158.

[71] Mizuno A, Oikawa K, Aoyagi M, Kajiwara H, Tamura H, Takano T. Examination of High-Torque Sandwich-Type Spherical Ultrasonic Motor Using with High-Power Multimode Annular Vibrating Stator. Actuators. 2018 Feb 27; 7(1):8. doi: 10.3390/act7010008. doi: 10.3390/act7010008.

[72] Ma Y, Choi M, Uchino K. Single-phase driven ultrasonic motor using two orthogonal bending modes of sandwiching piezo-ceramic plates. Rev Sci Instrum. 2016 Nov;87(11):115004. doi: 10.1063/1.4967857.

[73] Feng Z, Liang W, Ling J, Xiao X, Tan KK, Lee TH. Integral terminal sliding-mode-based adaptive integral backstepping control for precision motion of a piezoelectric ultrasonic motor. Mech Syst Signal Process. 2020 Oct; 144:106856. doi: 10.1016/j.ymssp.2020.106856.

[74] Chen C, Shi Y, Zhang J, Wang J. Novel linear piezoelectric motor for precision position stage. Chin J Mech Eng. 2016 Mar; 29(2):378-385. doi: 10.3901/CJME.2015.1216.149.

[75] Dong X, Hu M, Jin L, Xu Z, Jiang C. A standing wave ultrasonic stepping motor using open-loop control system. Ultrasonics. 2018 Jan; 82:327-330. doi: 10.1016/j.ultras.2017.09.014.

[76] Chen X, Li M, Zhang H, Lu Q, Lyu S. Improvement on the Structure Design of a Kind of Linear Piezoelectric Motor with Flexible Drive-Foot. Int J Precis Eng Manuf. 2020 Jan;21(1): 81-89.

[77] Kanada A, Mashimo T. Design and Experiments of Flexible Ultrasonic Motor Using a Coil Spring Slider. IEEEASME Trans Mechatron. 2020 Feb;25(1):468-476.

[78] Sato Y, Kanada A, Mashimo T. Self-Sensing and Feedback Control for a Twin Coil Spring-Based Flexible Ultrasonic Motor. IEEE Robot Autom. Lett. 2020;1-1.

[79] Dabbagh V, Sarhan AAD, Akbari J, Mardi NA. Design and experimental evaluation of a precise and compact tubular ultrasonic motor driven by a single-phase source. Precis Eng. 2017 Apr;48:172-180.

[80] Dabbagh V, Sarhan AAD, Akbari J, Mardi NA. Design and manufacturing of ultrasonic motor with in-plane and out-of-plane bending vibration modes of rectangular plate with large contact area. Measurement. 2017 Oct; 109: 425-431.

[81] Mažeika D, Čeponis A, Makutėnienė D. A Cylinder-Type Multimodal Traveling Wave Piezoelectric Actuator. Appl Sci. 2020 Apr 1; 10(7):2396.

[82] Čeponis A, Mažeika D, Vasiljev P. Flat Cross-Shaped Piezoelectric Rotary Motor. Appl Sci. 2020 Jul 21; 10(14):5022.

[83] Wang G, Xu W, Gao S, Yang B, Lu G. An energy harvesting type ultrasonic motor. Ultrasonics. 2017 Mar; 75:22-27.

[84] Wang G, Zhao Z, Tan J, Cui S, Wu H. A novel multifunctional piezoelectric composite device for mechatronics systems by using one single PZT ring. Smart Mater Struct. 2020 May 1; 29(5):055027.

[85] Tanoue Y, Morita T. Opposing preloads type ultrasonic linear motor with quadruped stator. Sens Actuators Phys. 2020 Jan; 301:111764.

[86] Tanoue Y, Morita T. Rod drive type ultrasonic linear motor with quadruped stator. Jpn J Appl Phys. 2020 Jul 1; 59(SK): SKKD13.

[87] Cheon S-K, Jeong S-S, Ha Y-W, Lee B-H, Park J-K, Park T-G. Driving characteristics of an ultrasonic rotary motor consisting of four-line contact type stators. Ceram Int. 2015 Jul; 41:S618–S624.

[88] Peng T, Wu X, Liang X, Shi H, Luo F. Investigation of a rotary ultrasonic motor using a longitudinal vibrator and spiral fin rotor. Ultrasonics. 2015 Aug; 61:157-161.

[89] Mohd Romlay FR, Wan Yusoff WA, Mat Piah KA. Increasing the efficiency of traveling wave ultrasonic motor by modifying the stator geometry. Ultrasonics. 2016 Jan; 64:177-185.

[90] Le AY, Mills JK, Benhabib B. A design methodology for linear ultrasonic motors under contact-based operating conditions. J Intell Mater Syst Struct. 2016 Jan; 27(1):39-50.

[91] Pan Q, Huang F, Chen J, He LG, Li W, Feng Z. High-Speed Low-Friction Piezoelectric Motors Based On Centrifugal Force. IEEE Trans Ind Electron. 2017 Mar; 64(3):2158-2167.

[92] Yang L, Zhu X, Di S. A type of dual-rotor hybrid ultrasonic motor based on vibration of four side panels. J Intell Mater Syst Struct. 2017 Aug; 28(14):1916-1924.

[93] Liu J, Jin J, Ji R, Chen D. A novel modal-independent ultrasonic motor with dual stator. Ultrasonics. 2017 Apr; 76:177-182.

[94] Li J, Liu S, Zhou N, Yu A, Cui Y, Chen P. A traveling wave ultrasonic motor with a metal/polymer-matrix

material compound stator. Smart Mater Struct. 2018 Jan 1; 27(1):015027.

[95] Sanikhani H, Akbari J. Design and analysis of an elliptical-shaped linear ultrasonic motor. Sens Actuators Phys. 2018 Aug; 278:67-77.

[96] Sun D, Tang Y, Wang J, Wang X. A novel fixable cylindrical ultrasonic motor. Adv Mech Eng. 2019 Mar; 11(3):168781401982998.

[97] Chu X, Zhang M, Yuan S, Zheng X. An Eight-Zonal Piezoelectric Tube-Type Threaded Ultrasonic Motor Based on Second-Order Bending Mode. Appl Sci. 2019 May 16; 9(10):2018.

[98] Borodinas S, Vasiljev P, Mazeika D, Bareikis R, Yang Y. Design optimization of double ring rotary type ultrasonic motor. Sens Actuators Phys. 2019 Jul; 293:160-166.

[99] Yang C-P, Xie K-J, Chang J-Y. Design and simulation of an ultrasonic linear motor with dual piezoelectric actuators. Microsyst Technol. 2020 Jan; 26(1): 71-78.

[100] Hirano T, Aoyagi M, Kajiwara H, Tamura H, Takano T. Development of rotary-type noncontact-synchronous ultrasonic motor. Jpn J Appl Phys. 2019 Jul 1;58(SG): SGGD09.

[101] Xu D, Zhang X, Zhao L, Yu S. A Novel Rotary Ultrasonic Motor Using the Longitudinal Vibration Mode. IEEE Access. 2019; 7:135650-135655.

[102] Wu J, Mizuno Y, Nakamura K. A traveling-wave ultrasonic motor utilizing a ring-shaped alumina/PZT vibrator. Smart Mater Struct. 2019 Dec 1; 28(12):125017.

[103] Liu J, Wang S, Wang Z, Gao N, Zhang D. Estimation and elimination of eigenvalue splitting and vibration instability of ring-shaped periodic structure subjected to three-axis angular velocity components. Meccanica. 2019 Dec; 54(15):2539-2563.

[104] Wang J, Wu Y, Yang Z, Zhao X, Chen X, Zhao D. Design and experimental performance of a low-frequency piezoelectric motor based on inertia drive. Sens Actuators Phys. 2020 Apr; 304:111854.

[105] Wang Y-J, Chen Y-C, Shen S-C. Design and analysis of a standing-wave trapezoidal ultrasonic linear motor. J Intell Mater Syst Struct. 2015 Nov; 26(17):2295-2303.

[106] YANG Y. MULTI-SPOKE-TYPE ULTRASONIC MOTOR. Nanjing, Jiangsu; US 2020/0007052 A1, 2020.

[107] Mathias R. PIEZO MOTOR. Karlsruhe (DE); US 2020/0038913 A1, 2020.

[108] YANG H-P. ULTRASONIC LINEAR ACTUATION DEVICE. Hsinchu City; US 2020/0169189 A1, 2020.

[109] He S, Shi S, Zhang Y, Chen W. Design and Experimental Research on a Deep-Sea Resonant Linear Ultrasonic Motor. IEEE Access. 2018; 6:57249-57256.

[110] Nakajima S, Kajiwara H, Aoyagi M, Tamura H, Takano T. Study on spherical stator for multidegree-of-freedom ultrasonic motor. Jpn J Appl Phys. 2016 Jul 1; 55(7S1):07KE18.

[111] Chen Z, Chen Y, Zhou T. A hollow cylindrical linear nut-type ultrasonic motor. Mechanics. 2017 Jan 20; 22(6):546-552.

[112] Li X, Yao Z, Wu R. Modelling and sticking motion analysis of a vibro-impact system in linear ultrasonic motors. Int J Mech Sci. 2015 Sep; 100:23-31.

[113] Li X, Yao Z, Zhou S, Lv Q, Liu Z. Dynamic modelling and characteristics

analysis of a modal-independent linear ultrasonic motor. Ultrasonics. 2016 Dec; 72:117-127.

[114] Li X, Yao Z, He Y, Dai S. Modelling and experimental investigation of thermal–mechanical–electric coupling dynamics in a standing wave ultrasonic motor. Smart Mater Struct. 2017 Sep 1;26(9):095044.

[115] Li X, Chen Z, Yao Z. Contact analysis and performance evaluation of standing-wave linear ultrasonic motors via a physics-based contact model. Smart Mater Struct. 2019 Jan 1;28(1):015032.

[116] Li X, Kan C, Cheng Y, Chen Z, Ren T. Performance evaluation of a bimodal standing-wave ultrasonic motor considering nonlinear electroelasticity: Modelling and experimental validation. Mech Syst Signal Process. 2020 Jul; 141:106475.

[117] Lv Q, Yao Z, Li X. Contact analysis and experimental investigation of a linear ultrasonic motor. Ultrasonics. 2017 Nov; 81:32-38.

[118] Li S, Li D, Yang M, Cao W. Parameters identification and contact analysis of traveling wave ultrasonic motor based on measured force and feedback voltage. Sens Actuators Phys. 2018 Dec; 284:201-208.

[119] Jiang C, Dong X, Jin L, Lu D. Contact Modelling and Performance Evaluation of a Radial Standing Wave Ultrasonic Motor. Math Probl Eng. 2019 May 19; 2019:1-10.

[120] Zhu Y, Yang T, Fang Z, Shiyang L, Cunyue L, Yang M. Contact modelling for control design of traveling wave ultrasonic motors. Sens Actuators Phys. 2020 Aug; 310:112037.

[121] Li J, Liu S, Qu J, Cui Y, Liu Y. A contact model of traveling-wave ultrasonic motors considering preload and load torque effects. Int J Appl Electromagn Mech. 2018 Feb 27; 56(2):151-164.

[122] Lv Q, Yao Z, Li X. Modelling and experimental validation of a linear ultrasonic motor considering rough surface contact. Smart Mater Struct. 2017 Apr 1; 26(4):045023.

[123] He Y, Yao Z, Dai S, Zhang B. Hybrid simulation for dynamic responses and performance estimation of linear ultrasonic motors. Int J Mech Sci. 2019 Apr; 153-154:219-229.

[124] Dai S, Yao Z, Zhou L, He Y. Modelling and analysis of a linear ultrasonic motor: consider the slider movement characteristic. Smart Mater Struct. 2019 Oct 1;28(10):105028.

[125] Zhou L, Yao Z, Li X, Dai S. Modelling and verification of thermal-mechanical-electric coupling dynamics of a V-shape linear ultrasonic motor. Sens Actuators Phys. 2019 Oct; 298:111580.

[126] Li X, Yao Z, Li R, Wu D. Dynamics modelling and control of a V-shaped ultrasonic motor with two Langevin-type transducers. Smart Mater Struct. 2020 Feb 1; 29(2):025018.

[127] Li C, Min R, Lu C. Note: Performance estimation of a rotary ultrasonic motor based on two-dimensional analytical model. Rev Sci Instrum. 2018 Oct; 89(10):106104.

[128] Renteria Marquez IA, Bolborici V. A dynamic model of the piezoelectric traveling wave rotary ultrasonic motor stator with the finite volume method. Ultrasonics. 2017 May; 77:69-78.

[129] Renteria-Marquez IA, Renteria-Marquez A, Tseng BTL. A novel contact model of piezoelectric traveling wave rotary ultrasonic motors with the finite volume method. Ultrasonics. 2018 Nov; 90:5-17.

[130] Mashimo T, Terashima K. Experimental Verification of Elliptical Motion Model in Traveling Wave Ultrasonic Motors. IEEEASME Trans Mechatron. 2015 Dec; 20(6):2699-2707.

[131] Mashimo T, Terashima K. Dynamic analysis of an ultrasonic motor using point contact model. Sens Actuators Phys. 2015 Sep; 233:15-21.

[132] Wang L, Hofmann V, Bai F, Jin J, Twiefel J. A novel additive manufactured three-dimensional piezoelectric transducer: Systematic modelling and experimental validation. Mech Syst Signal Process. 2019 Jan; 114:346-365.

[133] Wang L, Hofmann V, Bai F, Jin J, Twiefel J. Modelling of coupled longitudinal and bending vibrations in a sandwich type piezoelectric transducer utilizing the transfer matrix method. Mech Syst Signal Process. 2018 Aug; 108:216-237.

[134] Yang X, Liu Y, Chen W, Zhao X. New Excitation Method for Sandwich Transducer Using Bending Composite Vibrations: Modelling, Simulation, and Experimental Evaluation. IEEE Trans Ind Electron. 2018 Jun; 65(6): 4889-4896.

[135] Deng Y, Zhao G, Yi X, Xiao W. Contact modelling and input-voltage-region based parametric identification for speed control of a standing wave linear ultrasonic motor. Sens Actuators Phys. 2019 Aug; 295:456-468.

[136] Deng Y, Zhao G, Yi X, Xiao W. Comprehensively modelling and parametric identification for speed prediction of L1B2 ultrasonic motor. J Phys Conf Ser. 2020 Jan; 1449:012007.

[137] Zhang D, Wang S, Xiu J. Piezoelectric parametric effects on wave vibration and contact mechanics of traveling wave ultrasonic motor. Ultrasonics. 2017 Nov; 81:118-126.

[138] Zhang D, Wang S, Xiu J. Distorted wave response of ultrasonic annular stator incorporating non-uniform geometry. Wave Motion. 2017 Jan; 68:43-55.

[139] Velázquez R. An analytical and experimental study of ultrasonic linear motors. Teh Vjesn-Tech Gaz. 2015 Aug 30; 22(4):1057-1063.

[140] Nitan I, Milici LD, Milici MR, Cernomazu D. Elements of Modeling and Simulation for Piezoelectric Motors. Key Eng Mater. 2015 Aug; 660:323-329.

[141] Shi W, Zhao H, Ma J, Yao Y. Optimal working frequency of ultrasonic motors. Ultrasonics. 2016 Aug;70:38-44.

[142] Numerical Prediction of Stator Diameter Effect on the Output Torque of Ultrasonic Traveling-wave Motor, using Finite Elements Simulation. Int J Eng. V, vol.29, 2016 May. Available from: http://www.ije.ir/Vol29/No5/B/16.pdf

[143] An D, Ning Q, Huang W, Xue H, Zhang J. Effect of Damping Factor Variation on Eigenfrequency Drift for Ultrasonic Motors. In: Yu H, Liu J, Liu L, Ju Z, Liu Y, Zhou D, editors. Intelligent Robotics and Applications Springer International Publishing; 2019 [cited 2020 Aug 19]. p. 285-91. (Lecture Notes in Computer Science; vol. 11741). Available from: http://link.springer.com/10.1007/978-3-030-27532-7_26

[144] Ran L, Hu j, W. Zhou, . Study on the Mode Matching Issue of MEMS Traveling Wave Ultrasonic Motor. Mech Eng Technol. 2020;09(02):100-107.

[145] Wang C. How to Determine the Size of Longitudinal-flexural Mode Linear Ultrasonic Motor. J Phys Conf Ser. 2020 Jun; 1549:032062.

[146] Yin Z, Dai C, Cao Z, Li W, Chen Z, Li C. Modal analysis and moving performance of a single-mode linear

ultrasonic motor. Ultrasonics. 2020 Dec; 108:106216.

[147] Huang J, Sun D. Performance Analysis of a Travelling-Wave Ultrasonic Motor under Impact Load. Micromachines. 2020 Jul 16; 11(7):689.

[148] Chen N, Fan D. A teeth-discretized electromechanical model of a traveling-wave ultrasonic motor. Mech Sci. 2020 Jul 9; 11(2):257-266.

[149] Li X, Yao Z, Mi Y, Lin X, Liang C, Wu D. Modelling, analysis and suppression of current harmonics of Langevin-type ultrasonic motors under high voltage. Precis Eng. 2020 Jul; 64:177-187.

[150] Carvalho PA, Tang H, Razavi P, Pooladvand K, Castro WC, Gandomi KY, et al. Study of MRI Compatible Piezoelectric Motors by Finite Element Modeling and High-Speed Digital Holography. Advancements in Optical Methods & Digital Image Correlation in Experimental Mechanics, Volume 3, Springer International Publishing; 2020. p. 105-12. (Conference Proceedings of the Society for Experimental Mechanics Series). Available from: http://link.springer.com/10.1007/978-3-030-30009-8_16

[151] Liu Q, Huo X, Shi W, Zhao H. Experimental Modeling of Rotary Traveling-Wave Ultrasonic Motor. In: Wang R, Chen Z, Zhang W, Zhu Q, editors. Proceedings of the 11th International Conference on Modelling, Identification and Control (ICMIC2019) [Internet]. Singapore: Springer Singapore; 2020 [cited 2020 Aug 19]. p. 875-85. (Lecture Notes in Electrical Engineering; vol. 582). Available from: http://link.springer.com/10.1007/978-981-15-0474-7_82

[152] Mustafa A, Morita T. Modelling of Preload Controllable Rotary Ultrasonic Motors. Proceedings of JSPE Semestrial Meeting, 2019, Volume 2019S, 2019 JSPE Spring Conference, Pages 750-751, Released September 04, 2019, https://doi.org/10.11522/pscjspe.2019S.0_750.

[153] Costa Conrado A. Modelling the radially polarised annular stator of a piezoelectric travelling wave ultrasonic motor based on the shear effect. J Intell Mater Syst Struct. 2019 May; 30(8): 1225-1238.

[154] Ren W, Yang L, Ma C, Li X, Zhang J. Output performance simulation and contact analysis of traveling wave rotary ultrasonic motor based on ADINA. Comput Struct. 2019 May; 216:15-25.

[155] Zhang B, Wang Z, Wang T. Development of two-axis non-magnetic turntable based on ultrasonic motor. Adv Mech Eng. 2019 Mar; 11(3):168781401982858.

[156] Chen H, Chen C, Wang J, Shi M. Performance analysis and experimental study of traveling wave type rotary ultrasonic motor in high-rotation environment. Rev Sci Instrum. 2018 Nov; 89(11):115004.

[157] Liu W, Zhou M, Ruan X, Fu X. Non-smooth model and numerical analysis of a friction driven structure for piezoelectric motors. Int J Non-Linear Mech. 2017 May; 91:140-150.

[158] Latrèche S, Mostefai M, Meddad M, Eddiai A, Sahraoui B, Khemliche M, et al. Modelling and diagnostic of an ultrasonic piezoelectric actuator. Mol Cryst Liq Cryst. 2016 Mar 23; 628(1):23-40.

[159] Ri C-S, Kim M-J, Kim C-S, Im S-J. Study on the vibration displacement distribution of a circular ultrasonic motor stator. Ultrasonics. 2015 May; 59:59-63.

[160] Nguyen MHT, Liang W, Teo C-S, Tan K-K. Piecewise affine modeling and

compensation in motion of linear ultrasonic actuators. Mechatronics. 2015 Apr; 27:20-27.

[161] Song J, Liu X, Zhao G, Ding Q, Qiu J. Effect of surface roughness and reciprocating time on the tribological properties of the polyimide composites: Tribology of polyimide composites. Polym Eng Sci. 2019 Mar; 59(3):483-489.

[162] Liu X, Qiu J, Zhao G. Improved energy conversion efficiency of the ultrasonic motor with surface texture. Ind Lubr Tribol. 2018 Nov 19; 70(9):1729-1736.

[163] Song J, Yu Y, Zhao G, Qiu J, Ding Q. Comparative study of tribological properties of insulated and conductive polyimide composites. Friction. 2020 Jun; 8(3):507-516.

[164] Liu X, Song J, Chen H, Zhao G, Qiu J, Ding Q. Enhanced transfer efficiency of ultrasonic motors with polyimide based frictional materials and surface texture. Sens Actuators Phys. 2019 Aug; 295:671-677.

[165] Li J, Zhou N, Yu A, Cui Y. Tribological Behavior of CF/PTFE Composite and Anodized Al-Rotor in Traveling Wave Ultrasonic Motors. Tribol Lett. 2017 Mar; 65(1):4.

[166] Li J, Zeng S, Liu S, Zhou N, Qing T. Tribological properties of textured stator and PTFE-based material in travelling wave ultrasonic motors. Friction. 2020 Apr; 8(2):301-310.

[167] Zeng S, Li J, Zhou N, Zhang J, Yu A, He H. Improving the wear resistance of PTFE-based friction material used in ultrasonic motors by laser surface texturing. Tribol Int. 2020 Jan; 141:105910.

[168] Ishii T, Yamawaki H, Ohuchi H. An Ultrasonic Motor Using Thrust Bearing with Dimple Structure on the Friction Surface. Key Eng Mater. 2015 Jun; 649:54-59.

[169] Song F, Yang Z, Zhao G, Wang Q, Zhang X, Wang T. Tribological performance of filled PTFE-based friction material for ultrasonic motor under different temperature and vacuum degrees: ARTICLE. J Appl Polym Sci. 2017 Oct 15; 134(39):45358.

[170] Tribological behavior prediction of friction materials for ultrasonic motors using Monte Carlo-based artificial neural network.pdf.

[171] Li S, Zhang N, Yang Z, Li X, Zhao G, Wang T, et al. Tailoring friction interface with surface texture for high-performance ultrasonic motor friction materials. Tribol Int. 2019 Aug; 136:412-420.

[172] Yu T-H. Friction Layer Analysis of a Surface Acoustic Wave Motor. J Tribol. 2020 Sep 1; 142(9):091202.

[173] Qiu W, Mizuno Y, Nakamura K. Tribological performance of ceramics in lubricated ultrasonic motors. Wear. 2016 Apr; 352-353:188-195.

[174] Qiu W, Mizuno Y, Adachi K, Nakamura K. Ultrasonic motor performance influenced by lubricant properties. Sens Actuators Phys. 2018 Oct; 282:183-191.

[175] Wang Q, Song F, Zhang X, Zhao G, Wang T. Impact of fillers and counter face topography on wear behaviour of PTFE polymers for ultrasonic motor: ARTICLE. J Appl Polym Sci. 2017 May 19; 134(19). Available from: http://doi.wiley.com/10.1002/app.44835

[176] Zhao G, Wu C, Zhang L, Song J, Ding Q. Friction and wear behavior of PI and PTFE composites for ultrasonic motors. Polym Adv Technol. 2018 May; 29(5):1487-1496.

[177] An G, Li H. Degradation State Identification of Cracked Ultrasonic

Motor by Means of Fault Feature Extraction Method. Shock Vib. 2019 Mar 28; 2019:1-13.

[178] An G, Song K, Li R, Sun H, Li H. Degradation feature extraction method for piezoelectric ceramic of ultrasonic motor based on DCT-SV cross entropy. J Vibroengineering. 2019 Sep 30; 21(6):1651-1664.

[179] An G, Li R, Song K, Sun H, Xue Z, Li Z. Degradation State Identification Method for Piezoelectric Ceramic of Ultrasonic Motor Based on Segmented Fractal Dimension and Sparse Representation. J Test Eval. 2021 Sep 1; 49(5):20190800.

[180] An G, Li R, Song K, Sun H, Li H. Degradation State Identification for Ceramic in Ultrasonic Motor Based on Morphological Boundary Span Analysis. J Fail Anal Prev. 2019 Jun; 19(3): 761-770.

[181] Li J, Qu J, Zhang Y. Wear properties of brass and PTFE–matrix composite in traveling wave ultrasonic motors. Wear. 2015 Sep; 338-339:385-393.

[182] Zhang Y, Qu J, Li J. Friction and wear behaviour of linear standing-wave ultrasonic motors with V-shape transducers. Tribol Int. 2016 Mar; 95:95-108.

[183] Zhang Y, Qu J, Wang H. Wear Characteristics of Metallic Counterparts under Elliptical-Locus Ultrasonic Vibration. Appl Sci. 2016 Oct 11; 6(10):289.

[184] Chen X, Lu Q, Huang W, Wang Y. Working Mechanism of Nonresonance Friction in Driving Linear Piezoelectric Motors with Rigid Shaking Beam. Math Probl Eng. 2018 Nov 28;2018:1-10.

[185] Li X, Yao Z, Wu R., modelling and analysis of stick-slip motion in a linear piezoelectric ultrasonic motor considering ultrasonic oscillation effect. Int J Mech Sci. 2016 Mar; 107:215-224.

[186] Li X, Yao Z, Lv Q, Liu Z. modelling stick-slip-separation dynamics in a bimodal standing wave ultrasonic motor. J Sound Vib. 2016 Nov; 382: 140-157.

[187] Yang L, Wang F, Zhang J, Ren W. Remaining useful life prediction of ultrasonic motor based on Elman neural network with improved particle swarm optimization. Measurement. 2019 Sep; 143:27-38.

[188] Zhang L, Zheng H, Huang S, Zhang W, Li F, Liu D. A life test of ultrasonic motors under different torque loads and the analysis of the characteristics of wearing surfaces. Proc Inst Mech Eng Part J J Eng Tribol. 2020 May; 234(5):770-777.

[189] Kong D, Kurosawa MK. Excitation of Rayleigh wave with sapphire/$LiNbO_3$ mechanical integration for surface acoustic wave motor. Jpn J Appl Phys. 2018 Jul 1 ;57(7S1):07LE07.

[190] Padgurskas J, Rukuiža R, Bansevičius R, Jūrėnas V, Bubulis A. Impact of the tribological characteristics on the dynamics of the ultrasonic piezoelectric motor. Mechanics. 2015 Mar 6; 21(1):51-55.

[191] Zhang Y, Fu Y, Hua X, Quan L, Qu J. Wear debris of friction materials for linear standing-wave ultrasonic motors: Theory and experiments. Wear. 2020 May; 448-449:203216.

[192] Zhang Q, Li C, Zhang J, Zhang J. Smooth adaptive sliding mode vibration control of a flexible parallel manipulator with multiple smart linkages in modal space. J Sound Vib. 2017 Dec; 411:1-19.

[193] Pan S, Wu Y, Zhang J, Zhou S, Zhu H. modelling and control of a 2-degree-of-freedom gyro-stabilized platform driven by ultrasonic motors.

J Intell Mater Syst Struct. 2018 Jul; 29(11):2324-2332.

[194] Lu S, Jingzhuo S. Nonlinear Hammerstein model of ultrasonic motor for position control using differential evolution algorithm. Ultrasonics. 2019 Apr; 94:20-27.

[195] Jingzhuo S, Wenwen H, Ying Z. T–S Fuzzy Control of Travelling-Wave Ultrasonic Motor. J Control Autom Electr Syst. 2020 Apr; 31(2):319-328.

[196] Song L, Jingzhuo S. Novel Generalized Predictive Iterative Learning Speed Controller for Ultrasonic Motors. IEEE Access. 2020; 8:29344-29353.

[197] Lu S, Jingzhuo S, Ying Z. Secant iterative learning control of ultrasonic motor. ISA Trans. 2020 Aug;103: 343-354.

[198] Jingzhuo S, Huang W. Predictive Iterative Learning Speed Control With On-Line Identification for Ultrasonic Motor. IEEE Access. 2020; 8:78202-78212.

[199] Mo J-S, Qiu Z-C, Wei J-Y, Zhang X-M. Adaptive positioning control of an ultrasonic linear motor system. Robot Comput-Integr Manuf. 2017 Apr;44: 156-173.

[200] Pan S. Robust control of gyro stabilized platform driven by ultrasonic motor. Sens Actuators Phys. 2017 Jul; 261:280-287.

[201] Song P, Xu Z-F, Huang W-Q. Digital Internal Module Controller Shaped by Sensitivity Function of Platform Driven by Ultrasonic Motor. Int J Appl Phys Math. 2018; 8(3):45-52.

[202] Zeng W, Pan S, Chen L, Xu Z, Xiao Z, Zhang J. Research on ultra-low speed driving method of traveling wave ultrasonic motor for CMG. Ultrasonics. 2020 Apr; 103:106088.

[203] Shi W, Zhao H, Ma J, Yao Y. An Optimum-Frequency Tracking Scheme for Ultrasonic Motor. IEEE Trans Ind Electron. 2017 Jun; 64(6):4413-4422.

[204] Kebbab FZ, Jabri D, Belkhiat DEC, Belkhiat S. Frequency Speed Control of Rotary Travelling Wave Ultrasonic Motor Using Fuzzy Controller. 2018; 8(4):6.

[205] Fang Z, Yang T, Zhu Y, Li S, Yang M. Velocity Control of Traveling-Wave Ultrasonic Motors Based on Stator Vibration Amplitude. Sensors. 2019 Dec 3; 19(23):5326.

[206] Lee D-J, Lee S-K. Ultraprecision XY stage using a hybrid bolt-clamped Langevin-type ultrasonic linear motor for continuous motion. Rev Sci Instrum. 2015 Jan; 86(1):015111.

[207] Tavallaei MA, Atashzar SF, Drangova M. Robust Motion Control of Ultrasonic Motors Under Temperature Disturbance. IEEE Trans Ind Electron. 2016 Apr;63(4): 2360-2368.

[208] Liang W, Ma J, Ng C, Ren Q, Huang S, Tan KK. Optimal and intelligent motion control scheme for an Ultrasonic-Motor-Driven X-Y stage. Mechatronics. 2019 May; 59:127-139.

[209] Lin Y, Shi Y, Zhang J, Wang F, Wu W, Sun H. Design and Control of a Piezoelectric Actuated Prostate Intervention Robotic System *. In: 2020 17th International Conference on Ubiquitous Robots (UR) [Internet]. Kyoto, Japan: IEEE; 2020 p. 175-180. Available from: https://ieeexplore.ieee.org/document/9144768/

[210] Li X, Yao Z, Zhou L, Zhou S. Dispersed operating time control of a mechanical switch actuated by an ultrasonic motor. J Vibroengineering. 2018 Feb; 20(1):321-331.

[211] Koc B. Method For Closed-Loop Motion Control For An Ultrasonic

Motor. Karlsruhe; US 2020/0204088 A1, 2020.

[212] Delibas B, Koc B. A method to realize low velocity movability and eliminate friction induced noise in piezoelectric ultrasonic motors. IEEEASME Trans Mechatron. 2020; 1-1.

[213] Hieu NT, Odomari S, Yoshida T, Senjyu T, Yona A, Thick VH. Digital Position Control Strategy of Traveling-wave Ultrasonic Motors. Automatika. 2014 Jan; 55(3):246-255.

[214] Uebayashi A. ULTRASONIC MOTOR, DRIVE CONTROL SYSTEM, OPTICAL APPARATUS, AND VIBRATOR. Tokyo; US 10,536,097 B2, 2020.

[215] Ho S-T, Jan S-J. A piezoelectric motor for precision positioning applications. Precis Eng. 2016 Jan; 43:285-293.

[216] Wen Z, He Q, Qiao G. An insight into the flexible drive mechanism in short cylinder ultrasonic piezoelectric vibrator. Rev Sci Instrum. 2020 May; 91(5):055003.

[217] Qu, Y., Zhang, Y. and Qu, J. (2016), Micro-Driving behaviour of carbon-fiber-reinforced epoxy resin for standing-wave ultrasonic motor. Polym. Compos., 37: 2152-2159. https://doi-org.proxy.lib.odu.edu/10.1002/pc.23394

[218] Brahim M, Bahri I, Bernard Y. Real time implementation of H-infinity and RST motion control of rotary traveling wave ultrasonic motor. Mechatronics. 2017 Jun; 44:14-23.

[219] Chen N, Zheng J, Jiang X, Fan S, Fan D. Analysis and control of micro-stepping characteristics of ultrasonic motor. Front Mech Eng [Internet]. 2020 Mar 23 [cited 2020 Aug 19]; Available from: http://link.springer.com/10.1007/s11465-019-0577-3

[220] Chen N, Fan S, Fan D. A Multi-parameter driving and measurement circuit of ultrasonic motor. IOP Conf Ser Mater Sci Eng. 2020 Mar 31; 768:062011.

[221] Mustafa A, Morita T. Dynamic preload control of traveling wave rotary ultrasonic motors for energy efficient operation. Jpn J Appl Phys. 2019 Jul; 58(SG):SGGD04.

[222] Yokozawa H, Twiefel J, Weinstein M, Morita T. Dynamic resonant frequency control of ultrasonic transducer for stabilizing resonant state in wide frequency band. Jpn J Appl Phys. 2017 Jul 1; 56(7S1):07JE08.

[223] Zhang Q, Li C, Zhang J, Zhang X. Synchronized motion control and precision positioning compensation of a 3-DOFs macro–micro parallel manipulator fully actuated by piezoelectric actuators. Smart Mater Struct. 2017 Nov 1; 26(11):115001.

[224] Ghenna S, Giraud F, Giraud-Audine C, Amberg M. Vector Control of Piezoelectric Transducers and Ultrasonic Actuators. IEEE Trans Ind Electron. 2018 Jun; 65(6):4880-4888.

[225] Li H, Tian X, Shen Z, Li K, Liu Y. A low-speed linear stage based on vibration trajectory control of a bending hybrid piezoelectric ultrasonic motor. Mech Syst Signal Process. 2019 Oct; 132:523-534.

[226] Yonemoto D, Yashiro D, Yubai K, Komada S. Design of force control system using tendon-driven mechanism including linear springs and ultrasonic motor. Electr Eng Jpn. 2018 Oct; 205(1):36-45.

[227] Li X, Yao Z, Wu D. A Novel Nonlinear Tuning Method for Sandwich Type Ultrasonic Motors Under High Voltage. In: 2019 IEEE International Ultrasonics Symposium (IUS)

[Internet]. Glasgow, United Kingdom: IEEE; 2019 [cited 2020 Aug 19]. p. 1682-4. Available from: https://ieeexplore.ieee.org/document/8925851/

[228] Wischnewskiy W., Method for Operating an Ultrasonic Motor. Karlsruhe; US 2020/0035904 A1, 2020.

[229] Wu J, Mizuno Y, Tabaru M, Nakamura K. Traveling wave ultrasonic motor using polymer-based vibrator. Jpn J Appl Phys. 2016 Jan 1; 55(1):018001.

[230] Wu J, Mizuno Y, Nakamura K. Ultrasonic motors with poly phenylene sulfide/alumina/PZT triple-layered vibrators. Sens Actuators Phys. 2018 Dec; 284:158-167.

[231] Weng C-M, Tsai C-C, Hong C-S, Chiang C-H, Chu S-Y, Lin C-C, et al. Effects of post-annealing on electrical properties of $CuF_2 \cdot xH_2O$-doped KNN ceramics for rotary-linear ultrasonic motors. Ceram Int. 2018 Oct; 44(14):16173-16180.

[232] Doshida Y, Tamura H, Tanaka S. High-power properties of crystal-oriented $(Sr,Ca)_2NaNb_5O_{15}$ piezoelectric ceramics and their application to ultrasonic motors. Jpn J Appl Phys. 2019 Jul 1; 58(SG):SGGA07.

[233] Yu YH, Zhao G, Song JF, Ding QJ. Mechanical and Tribological Properties of Polyimide Composites for Reducing Weight of Ultrasonic Motors. Key Eng Mater. 2019 Apr; 799:65-70.

[234] Ou W, Li S, Cao W, Yang M. A single-mode Mn-doped 0.27PIN-0.46PMN-0.27PT single-crystal ultrasonic motor. J Electroceramics. 2016 Dec; 37(1-4):121-126.

[235] Shi X, Huang W, Li F, Li Z, Xu Z, Jiang X, et al. Analysis on the anisotropic electromechanical properties of lead magnoniobate titanate single crystal for ring type ultrasonic motors. AIP Adv. 2016 Nov; 6(11):115017.

[236] Li S, Yang R, Wang T, Zhang X, Wang Q. Surface textured polyimide composites for improving conversion efficiency of ultrasonic motor. Tribol Int. 2020 Dec; 152:106489.

[237] Wu J, Mizuno Y, Nakamura K. Structural parameter study on polymer-based ultrasonic motor. Smart Mater Struct. 2017 Nov 1; 26(11):115022.

[238] Wu J, Mizuno Y, Nakamura K. Polymer-Based Ultrasonic Motors Utilizing High-Order Vibration Modes. IEEEASME Trans Mechatron. 2018 Apr; 23(2):788-799.

[239] Cao T, Li X, Wang B, Mi Y, Zhao G, Twiefel J, et al. Viscoelastic analytical model and design of polymer-based bimodal piezoelectric motor. Mech Syst Signal Process. 2020 Nov; 145:106960.

[240] Chen Y, Mei K, Wong C-M, Lin D, Chan H, Dai J. Ultrasonic Transducer Fabricated Using Lead-Free BFO-BTO+Mn Piezoelectric 1-3 Composite. Actuators. 2015 May 29; 4(2):127-134.

[241] Zhao K, Wang D, Wang Z, Jiang C, Abbas Z, Zheng Y, et al. Fabrication of piezoelectric thick-film stator using electrohydrodynamic jet printing for micro rotary ultrasonic motors. Ceram Int. 2020 Jul; S0272884220321076.

[242] Zhao Y, Yuan S, Chu X, Gao S, Zhong Z, Zhu C. Ultrasonic micro-motor with multilayer piezoelectric ceramic and chamfered driving tips. Rev Sci Instrum. 2016 Sep; 87(9):095108.

[243] Mustafa A, Morita T. Efficiency optimization of rotary ultrasonic motors using extremum seeking control with current feedback. Sens Actuators Phys. 2019 Apr; 289:26-33.

[244] Le AY, Mills JK, Benhabib B. Improved linear ultrasonic motor

performance with square-wave based driving-tip trajectory. Smart Mater Struct. 2015 Mar 1; 24(3):037003.

[245] Shi W, Zhao B, Qi X, Wang Y, Zhao H, Chen W, et al. Pseudo-Full-Bridge Inverter With Soft-Switching Capability for a Quarter-Phase Ultrasonic Motor. IEEE Trans Ind Electron. 2019 Jun; 66(6):4199-4208.

[246] Li C, Lu C, Ma Y. Magnetic field tuning characteristics of bimodal ultrasonic motor stator. J Shanghai Jiaotong Univ Sci. 2017 Apr; 22(2):129-132.

[247] Chen N, Zheng J, Fan D. Pre-Pressure Optimization for Ultrasonic Motors Based on Multi-Sensor Fusion. Sensors. 2020 Apr 8; 20(7):2096.

[248] Peng T, Wu X, Liang X, Shi H, Luo F. Investigation of a rotary ultrasonic motor using a longitudinal vibrator and spiral fin rotor. Ultrasonics. 2015 Aug; 61:157-161.

[249] Shi W, Zhao H, Ma J, Yao Y. Dead-Zone Compensation of an Ultrasonic Motor Using an Adaptive Dither. IEEE Trans Ind Electron. 2018 May; 65(5):3730-3739.

[250] An D, Huang W. Inherent mechanism of frequency drift affected by constraint conditions for rotary piezoelectric motors. Rev Sci Instrum. 2020 Mar 1; 91(3):035002.

[251] Zhu C, Chu X, Yuan S, Zhong Z, Zhao Y, Gao S. Development of an ultrasonic linear motor with ultra-positioning capability and four driving feet. Ultrasonics. 2016 Dec; 72:66-72.

[252] Li X, Huang Y, Zhou L. Integrated performance improvement for a bimodal linear ultrasonic motor using a dual-frequency asymmetric excitation method. Ultrasonics. 2020 Dec; 108:106224.

[253] Zhou X, Zhang Y, Zhang Q. A novel linear ultrasonic motor with characteristic of variable mode excitation. Ceram Int. 2017 Aug; 43:S64–S69.

[254] Dong X, Jiang C, Jin L, Xu Z, Yuan Y. Inherent Loss Analysis of Piezoelectrics in Radial Vibration and its Application in Ultrasonic Motor. IEEE Trans UFFC. 2020 Aug; 67(8):1632-1640.

[255] He L, Hao S, Zhao X, Dong Y, Li X, Chen J, et al. Resonant-type piezoelectric linear motor driven by harmonic synthesized mechanical square wave. Rev Sci Instrum. 2020 Mar 1;91(3):035005.

[256] Nishizawa U, Oohashi T, Toyama S. 1939. Evaluation of spherical ultrasonic motor for space in high temperature condition. ISSN. 2016; 18(2):11.

[257] Nishizawa U, Oohashi T, Toyama S. Evaluation of spherical ultrasonic motor for space in low temperature condition. J Vibroengineering. 2017 Nov 15; 19(7):5170-5181.

[258] Shi W, Zhao H, Zhao B, Qi X, Chen W, Tan J. Extended optimum frequency tracking scheme for ultrasonic motor. Ultrasonics. 2018 Nov; 90:63-70.

[259] Sunif MFM, Haron MNF, Romlay FRM. Heat Energy Modelling of Travelling Wave Piezoceramic Ultrasonic Motor. Abdul Karim SA, Zainuddin N, Yusof MH, Sa'ad N, editors. MATEC Web Conf. 2018; 225:01009.

[260] Li H, Chen W, Tian X, Liu J. An experiment study on temperature characteristics of a linear ultrasonic motor using longitudinal transducers. Ultrasonics. 2019 May; 95:6-12.

[261] Cheng L-P, Zhang S-Y, Xu X-D. Effects of Transverse Temperature

Gradient on the Rotor Velocity in an Ultrasonic Motor. Chin Phys Lett. 2016 Jan; 33(1):014301.

[262] Nakazono M, Kanda T, Yamaguchi D, Suzumori K, Noguchi Y. A study on temperature dependence of an ultrasonic motor for cryogenic environment. Jpn J Appl Phys. 2015 Jul 1; 54(7S1):07HE15.

[263] Lv Q, Yao Z, Zhou L, Pan L. Effect of temperature rise on characteristics of a standing wave ultrasonic motor. J Intell Mater Syst Struct. 2019 Apr; 30(6):855-868.

[264] Peng Y, Peng Y, Gu X, Wang J, Yu H. A review of long range piezoelectric motors using frequency leveraged method. Sens Actuators Phys. 2015 Nov; 235:240-255.

[265] Spanner K, Koc B. Piezoelectric Motors, an Overview. Actuators. 2016 Feb 26; 5(1):6.

[266] Peled G, Yasinov R, Karasikov N. Performance and Applications of L1B2 Ultrasonic Motors. Actuators. 2016 Jun 1; 5(2):15.

[267] Gao X, Yang J, Wu J, Xin X, Li Z, Yuan X, et al. Piezoelectric Actuators and Motors: Materials, Designs, and Applications. Adv Mater Technol. 2020 Jan; 5(1):1900716.

[268] Shokrollahi P, Drake JM, Goldenberg AA. Signal-to-noise ratio evaluation of magnetic resonance images in the presence of an ultrasonic motor. Biomed Eng OnLine. 2017 Dec; 16(1):45.

[269] Shokrollahi P, Drake J, Goldenberg A. Measuring the Temperature Increase of an Ultrasonic Motor in a 3-Tesla Magnetic Resonance Imaging System. Actuators. 2017 Jun 6; 6(2):20.

[270] Shokrollahi P, Drake JM, Goldenberg AA. Ultrasonic motor-induced geometric distortions in magnetic resonance images. Med Biol Eng Comput. 2018 Jan;56(1):61-70.

[271] Shokrollahi P, Drake J, Goldenberg A. Quantification of Force and Torque Applied by a High-Field Magnetic Resonance Imaging System on an Ultrasonic Motor for MRI-Guided Robot-Assisted Interventions. Actuators. 2017 Sep 30; 6(4):29.

[272] Shokrollahi P, Drake JM, Goldenberg AA. A study on observed ultrasonic motor-induced magnetic resonance imaging (MRI) artifacts. Biomed J. 2019 Apr; 42(2):116-123.

[273] Carvalho PAWG, Nycz CJ, Gandomi KY, Fischer GS. Demonstration and Experimental Validation of Plastic-Encased Resonant Ultrasonic Piezoelectric Actuator for Magnetic Resonance Imaging-Guided Surgical Robots. J Eng Sci Med Diagn Ther. 2020 Feb 1; 3(1):011002.

[274] Liu, Wenxuan & Yang, Zhiyong & Jiang, Shan & Feng, Di & Zhang, Daguang. (2019). Design and implementation of a new cable-driven robot for MRI-guided breast biopsy. The International Journal of Medical Robotics and Computer Assisted Surgery. 16. 10.1002/rcs.2063.

[275] Hachisuka S, Kaneko T, Morita T. Clarification of Muscle Fatigue Reducing Effect of Walking Assist Device Using Electromyography. In: 2020 IEEE 2nd Global Conference on Life Sciences and Technologies (LifeTech) . Kyoto, Japan: IEEE; 2020. p. 161-162. Available from: https://ieeexplore.ieee.org/document/9081014/

[276] Kaneko T, Orino Y, Hachisuka S, Morita T. Effective Assist of Hip-joint Support Ambient Walking Assistive System Using Ultrasonic Motors. In: 2020 IEEE 2nd Global Conference on Life Sciences and Technologies (LifeTech). Kyoto, Japan: IEEE; 2020

[cited 2020 Aug 19]. p. 275-6. Available from: https://ieeexplore.ieee.org/document/9081021/

[277] Chen X, Chen Z, Li X, Shan L, Sun W, Wang X, et al. A spiral motion piezoelectric micromotor for autofocus and auto zoom in a medical endoscope. Appl Phys Lett. 2016 Feb; 108(5): 052902.

[278] Zhu P, Peng H, Yang J, Mao T, Sheng J. A New Low-frequency Sonophoresis System Combined with Ultrasonic Motor and Transducer. Smart Mater Struct. 2018 Mar 1; 27(3):035021.

[279] Netzel KE. ULTRASONIC SURGICAL INSTRUMENT WITH TORQUE ASSIST FEATURE. Loveland, CO; US 10,582,944 B2, 2020.

[280] Shi S, Huang Z, Yang J, Liu Y, Chen W, Uchino K. Development of a compact ring type MDOF piezoelectric ultrasonic motor for humanoid eyeball orientation system. Sens Actuators Phys. 2018 Apr; 272:1-10.

[281] Lu X, Wang Z, Shen H, Zhao K, Pan T, Kong D, et al. A Novel Dual-Rotor Ultrasonic Motor for Underwater Propulsion. Appl Sci. 2019 Dec 19; 10(1):31.

[282] Toyama S, Nishizawa U, Deep-sea Drone with Spherical Ultrasonic Motors. Tokyo University of Agriculture and Technology, Koganei, Tokyo, Japan, Int J Model Optim. 2019 Dec; 348-351.

[283] Pan S, Xu Z, Zhao C. A novel single-gimbal control moment gyroscope driven by an ultrasonic motor. Adv Mech Eng. 2019 Apr; 11(4):168781401984438.

[284] Xu Z, Pan S, Chen L, Di S, Huang W. A continuously variable beam expander driven by ultrasonic motors. Rev Sci Instrum. 2019 Sep; 90(9): 096107.

[285] Wang, Zhuo & WANG, Xin-tong & Wang, Tao & MA, Hong-wen. (2020). Research on Accuracy Analysis and Motion Control of Two-axis Non-magnetic Turntable Based on Ultrasonic Motor Journal. Mechanics. 26. 221-230. 10.5755/j01.mech.26. 3.23453.

[286] Wu R, Soutis C, Zhong S, Filippone A. A morphing aerofoil with highly controllable aerodynamic performance. Aeronaut J. 2017 Jan; 121(1235):54-72.

[287] Chen S, Wang L, Guo S, Zhao C, Tong M. A Bio-Inspired Flapping Wing Rotor of Variant Frequency Driven by Ultrasonic Motor. Appl Sci. 2020 Jan 6; 10(1):412.

[288] Wataru Y, Hiroyuki M, and Daizo I. 2019. ZeRONE: Safety Drone with Blade-Free Propulsion. In CHI Conference on Human Factors in Computing Systems Proceedings (CHI 2019), May 4-9, 2019, Glasgow, Scotland Uk. ACM, New York, NY, USA, Article 4, 8 pages. https://doi.org/10.1145/3290605.3300595

[289] Ibrahim A. S., "Development of a Standing Wave Tube Rotary Ultrasonic Piezoelectric Motor for LiDAR Systems.", Ph.D dissertation, University of Toronto, Toronto, 2018, Accessed on: November 16, 2020. Available:http://hdl.handle.net/1807/91982

[290] Chen C-C. An optical image stabilization using novel ultrasonic linear motor and fuzzy sliding-mode controller for portable digital camcorders. IEEE Trans Consum Electron. 2017 Nov;63(4):343-349.

[291] Hariri H, Bernard Y, Razek A. Dual piezoelectric beam robot: The effect of piezoelectric patches' positions. J Intell Mater Syst Struct. 2015 Dec; 26(18):2577-2590.

[292] Olsson P, Nysjo F, Carlbom IB, Johansson S. Comparison of Walking

and Traveling-Wave Piezoelectric Motors as Actuators in Kinesthetic Haptic Devices. IEEE Trans Haptics. 2016 Jul 1; 9(3):427-431.

[293] Geng R-R, Mills JK, Yao Z-Y. Design and analysis of a novel 3-DOF spatial parallel microgripper driven by LUMs. Robot Comput-Integr Manuf. 2016 Dec; 42:147-155.

[294] Li X, Zhou S. A novel piezoelectric actuator with a screw-coupled stator and rotor for driving an aperture. Smart Mater Struct. 2016 Mar 1; 25(3):035027.

[295] Ozeki S, Kurita K, Uehara C, Nakane N, Sato T, Takeuchi S. Analysis of coiled stator ultrasound motor: Fundamental study on analysis of wave propagation on acoustic waveguide for coiled stator. Jpn J Appl Phys. 2018 Jul 1; 57(7S1):07LB12.

[296] Lendraitis V, Gadišauskas T. Investigation of ultrasonic stepping motors and nanomanipulator for scanning probe microscopy. Mechanics. 2015 Sep 22; 21(4):334-338.

[297] Jian Y, Yao Z, Silberschmidt VV. Linear ultrasonic motor for absolute gravimeter. Ultrasonics. 2017 May; 77:88-94.

[298] Yamaoka Y, Funatsu K, Yoshidumi Y, Kubo A, Notsuka Y, Takahashi E. A compact scanning probe for photoacoustic microscopy using ultrasonic actuator stage. Jpn J Appl Phys. 2020 Mar 1; 59(3):030906.

[299] Kagermann, H.; Lukas, W.; Wahlster, W.: Industrie 4.0- Mit dem Internet der Dinge auf dem Weg zur 4. industriellen Revolution. In: VDI Nachrichten, Issue 13, (2011).

[300] Stock. T, Seliger G., Opportunities of Sustainable Manufacturing in Industry 4.0, Procedia CIRP, Volume 40, 2016, Pages 536-541, ISSN 2212-8271.

[301] Acatech: Umsetzungsempfehlungen für das Zukunftsprojekt Industrie 4.0 – Abschlussbericht des Arbeitskreises Industrie 4.0. acatech, (2013).

[302] Zhou K, Liu T and Zhou L, Industry 4.0: Towards future industrial opportunities and challenges, 2015 12th International Conference on Fuzzy Systems and Knowledge Discovery (FSKD), Zhangjiajie, 2015, pp. 2147-2152, doi: 10.1109/FSKD.2015.7382284.

[303] Industry 4.0: fourth industrial revolution guide to Industrie 4.0, 29-Sep-2020. [Internet]. Available: https://www.i-scoop.eu/industry-4-0/. [Accessed: 2020-Nov-25].

[304] Hunt H, Digital Twin: A Bridge Between Physical and Digital Worlds, Medium, 21-Nov-2019. [Internet]. Available: https://medium.com/neurisium/digital-twin-a-bridge-between-physical-and-digital-worlds-20a89791f072. [Accessed: 2020-Nov-25].

[305] Wang, Vincent Xi, and Wang, Lihui. Digital Twin-based WEEE Recycling, Recovery and Remanufacturing in the Background of Industry 4.0. International Journal of Production Research 57.12 (2018): 3892-902

[306] Mayr A et al., Electric Motor Production 4.0 – Application Potentials of Industry 4.0 Technologies in the Manufacturing of Electric Motors, 2018 8th International Electric Drives Production Conference (EDPC), Schweinfurt, Germany, 2018, pp. 1-13, doi:10.1109/EDPC.2018.8658294.

[307] Hastie T., Tibshirani R., and Friedman J. H., The elements of statistical learning: Data mining, inference, and prediction, 2nd ed. New York NY: Springer, 2009

[308] Herta C, Datenwissenschaften und maschinelles Lernen für die Industrie 4.0: Einsatzmöglichkeiten und

Fachkräftebedarf, in Industrie 4.0-Grundlagen und Anwendungen: Brachentreff der Berliner WIrtschaft und Industrie, S. Schäfer and C. Pinnow, Eds.: Beuth Verlag GmbH, 2015, pp. 27-35.

[309] Bauer D. et al., Handlungsfelder Additive Fertigungsverfahren, 1st ed. Düsseldorf: Verein Deutscher Ingenieure, 2016

[310] Reinhart G. et al., "Anwendungsfeld Automobilindustrie," in Handbuch Industrie 4.0: Geschäftsmodelle, Prozesse, Technik, G. Reinhart, Ed., München: Hanser, 2017, pp. 709-722

[311] Davies R. Industry 4.0: Digitalization for productivity and growth (2015). [Internet] Available from: https://www.europarl.europa.eu/thinktank/en/document.html?reference=EPRS_BRI%282015%29568337 [Accessed: 2020-11-25]

[312] Masood, Tariq, and Egger, Johannes. "Augmented Reality in Support of Industry 4.0—Implementation Challenges and Success Factors." Robotics and Computer-integrated Manufacturing 58 (2019): 181-195.